AF889650

Allemeersch

Polymerisation of Ethylene

Also of Interest

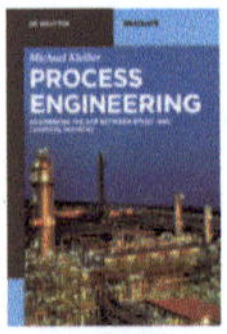

Process Engineering – Addressing the Gap between Studies and Chemical Industry
Michael Kleiber, 2015
ISBN 978-3-11-031209, e-ISBN 978-3-11-031211-9

Process Technology – An Introduction
André B. de Haan, 2015
ISBN 978-3-11-033671-9, e-ISBN 978-3-11-033672-6

Chemical Reaction Technology
Dimitry Yu. Murzin, 2015
ISBN 978-3-11-033643-6, e-ISBN 978-3-11-033644-3

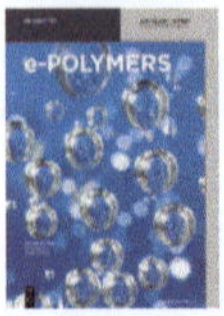

e-Polymers
Seema Agarwal, Andreas Greiner (Editor-in-Chief)
ISSN 1618-7229

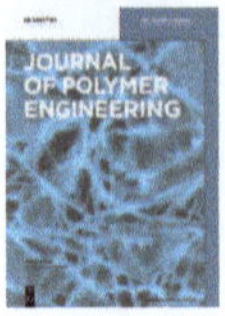

Journal of Polymer Engineering
Nino Grizzuti (Editor-in-Chief)
ISSN 2191-0340

Paul Allemeersch

Polymerisation of Ethylene

In Slurry Loop Reactors

DE GRUYTER

Author
Paul Allemeersch
Tessenderlo
Belgium

ISBN 978-3-11-029214-5
e-ISBN (PDF) 978-3-11-029219-0
e-ISBN (EPUB) 978-3-11-038886-2

Library of Congress Cataloging-in-Publication Data
A CIP catalog record for this book has been applied for at the Library of Congress.

Bibliographic information published by the Deutsche Nationalbibliothek
The Deutsche Nationalbibliothek lists this publication in the Deutsche Nationalbibliografie; detailed bibliographic data are available on the Internet at http://dnb.dnb.de.

© 2015 Walter de Gruyter GmbH, Berlin/Munich/Boston
Cover image: Fuse/Thinkstock
Typesetting: le-tex publishing services GmbH, Leipzig
Printing and binding: CPI books GmbH, Leck
♾ Printed on acid-free paper
Printed in Germany

www.degruyter.com

"Impression of a Loop Reactor" by the author

Preface

To engineers who operate a slurry loop reactor for the polymerisation of ethylene into high density polyethylene, this book presents all information necessary to understand the functioning of that reactor and operate it accordingly.

All discussions are based on experimental data from the operation of full scale commercial loop reactors, which makes it special and extremely useful. Too often is believed that the knowledge of a chemical process may only be gained from laboratories and pilot plants, and is forgotten to learn from the industrial plants, where instead of simulation, the real process is alive …

That this work is founded on full scale empirical data, does not mean that the material be not treated in depth. On the contrary, the author is convinced that the answer to a technical question should be formulated in most general terms first, and only then made solvable by a number of assumptions, clearly stated, and practicable by a number of empirical correlations. This approach leaves little room for errors, or doubts on the validity of the statements.

All necessary aspects of the loop reactor are thus treated, and then combined in a simple off-line reactor model, valid for all commercial types of catalysts. The book is completed by a detailed method to scale-up products from the laboratory directly to the full scale loop reactor, and by a discussion of the most important issues in the operation of two loop reactors in series.

It is tried to make the presented matter as comprehensible as possible. Thereto, all definitions are clearly indicated and described, and it is not avoided to introduce new definitions where appropriate. A separate chapter is dedicated to the building of data sets and the application of linear regression methods to obtain useful experimental information from full scale operations. The importance of certain conclusions to practical reactor operation is often mentioned.

If this book is concentrating on slurry loop reactors for high density polyethylene, it may also be an inspiration to engineers active in other processes, on how to learn from full scale operation and make this knowledge work.

For more than twenty years, the author was involved in the start-up of a new slurry loop reactor plant, the improvement of this process, and the development of its wide range of products. Solving the questions and problems met on this trajectory, eventually led to the systematic approach presented here.

The author feels grateful to the many excellent chemists, lab technicians and engineers who helped him to learn the process, and to organise a large series of full-scale test runs. They worked in Norway, Finland, England, Belgium and Sweden. He wishes especially to thank the operators, foremen, lab technicians and staff of the HDPE loop reactor plant in Beringen, Belgium, for their great work that made this book possible, and his colleague, Katleen Switten, for her encouragement, her careful reviewing of the text, and the numerous corrections and suggestions.

Tessenderlo, January 2015 Paul Allemeersch

Contents

1 Introduction to the Slurry Loop Process

The polymerisation products of ethylene are among the largest commodity plastics. Low density polyethylene (LDPE) is a highly branched polymer, generated from the monomer at conditions of extremely great pressure and high temperature. High density polyethylene (HDPE) is a linear polymer, and is obtained by catalysis, at moderate pressures and temperatures. The two types of polyethylene are so different in properties that they can hardly be said to be competitors.

Two main processes are known to produce HDPE. In the *gas phase process*, the polymer particles are formed on catalyst particles that are fluidized in a fluidized bed reactor by a gas stream carrying the monomer. In the *slurry loop process* the polymer grows on catalyst particles that are suspended in a liquid, wherein the monomer is dissolved, thus forming a slurry that is circulated in a recycle tubular reactor, commonly called *loop reactor*.

The slurry loop reactor is a long tube, about 600 *mm* in diameter and a few hundred metres long, of which the ends are connected to make it one closed loop. Its volume is of the order of 100 m^3. It is totally filled with the slurry, which is circulated through the loop by one single circulation pump. The larger part of the reactor consists of an even number of vertical tube sections, or *legs*, connected by short horizontal elbows at the top and the bottom.

The outer aspect of the reactor is that of a tubular reactor, one of the two main model reactors known in chemical reactor engineering, where the reagents are fed at one side of the tube and leave it at the other end. However, since the loop is closed, and the reactor contents are circulated through the loop at a high speed, the behaviour of the reactor is very close to that of the continuous stirred tank reactor, that other model known from reactor engineering textbooks.

Actually, the loop reactor can be regarded as a stirred tank, where the circulation pump has taken over the role of the agitator, and where the vessel has been lengthened to obtain more external cooling surface, and a higher production capacity. Indeed, the heat of reaction of the polymerisation of ethylene is quite high[1], and the cooling surface of the reactor is one of the limiting factors to its production capacity. Since a conventional vessel is rather spherical, the ratio of its external surface to its volume is inversely proportional to its diameter. In a loop reactor, the cooling of the reactor contents is effected by cooling water flowing through a cooling jacket, concentric with the tube, and so its external surface is simply proportional to its volume. When the reactor tube is made longer, both volume and cooling surface are increased to the same degree. Due to the fact that the slurry and the cooling water are flowing at very high speeds, high heat transfer coefficients are obtained at both sides of the reactor wall.

---

1 The heat of reaction of the polymerisation of ethylene is about 3720 *kJ* per *kg* of ethylene.

Three families of catalysts are used in the slurry loop process. They yield polymers of different molecular weight distributions (MWD), and thus allow to produce a very broad range of high density polyethylenes, of which many specific grades are used for specific applications. The reaction mechanisms of the three catalyst families are different, because their active species, that provoke the polymerisation reaction, are different. The active chemical of the *chromium oxide catalyst* is chromium(VI) oxide stabilised on a silica particle. The resulting MWD is very broad. The *Ziegler–Natta catalyst* is using a complex of $TiCl_4$ in an environment of $MgCl_2$. The resulting MWD is more narrow. The *metallocene catalyst*, where a Zr-metallocene group is the active chemical, produces the most narrow MWD. A chromium(VI) oxide catalyst cannot be realised without a silica support, but also the two other catalysts are most often supported on silica particles to facilitate their handling and feeding to the reactor. The size of those silica particles is of the order of 100 μm.
The catalysts are very sensitive to poisoning, especially by moisture. That is why every feed stream to the reactor is purified in an extensive *purification section*, where, on a series of adsorber beds, the impurity levels are reduced to the order of 0.1 *ppm*.

The best way to explain how the reactor functions, is by telling how it is started up. In a first step, the loop reactor is filled with the *diluent*. That is a liquid wherein the reaction ingredients are dissolved, but wherein the solubility of the polymer should be minimal. The most common diluent is isobutane, and also propane is used. The diluent is in the liquid state at the prevailing reactor conditions, viz. temperatures around 100 °C, and pressures around 40 *barg*.
The reactor is one stage in a greater loop described by this diluent in its course through the process, Figure 1.1, where it is passing a purification step before being fed to the reactor, is flashed off upon leaving the reactor, compressed and condensed, purified in the recovery and purification sections, and finally recycled to the reactor. By the filling of the reactor is meant that this circulation of diluent through the process is set up, and that the diluent in the reactor is at the operating temperature and pressure.

Then the required concentrations of monomer, and, if necessary, comonomer and hydrogen are established in the loop reactor. The condensed liquid, entering the *recovery section*, will then contain, besides the diluent, also concentrations of ethylene, comonomer and hydrogen. This stream is split up in its components in the recovery section by distillation, as far as is necessary to purify them and control their quantities that are all recycled to the reactor. Only two small purge streams, of lights and heavies, are taken off in the recovery section, which means that most of the liquid is recycled.

The reaction is started by the feeding of the catalyst. Polymer will be formed on the catalyst particles, and so a polymer-diluent slurry will be established. The thickening slurry is allowed to reach its steady state concentration, at some 20 *vol*%. This concentration is controlled by the feed flow of the recycled diluent, and by the number of settling legs in service.
The *settling legs* are vertical tubes suspended to the bottom of the reactor, wherein the slurry is allowed to settle and reach a higher concentration of solids. Small quantities

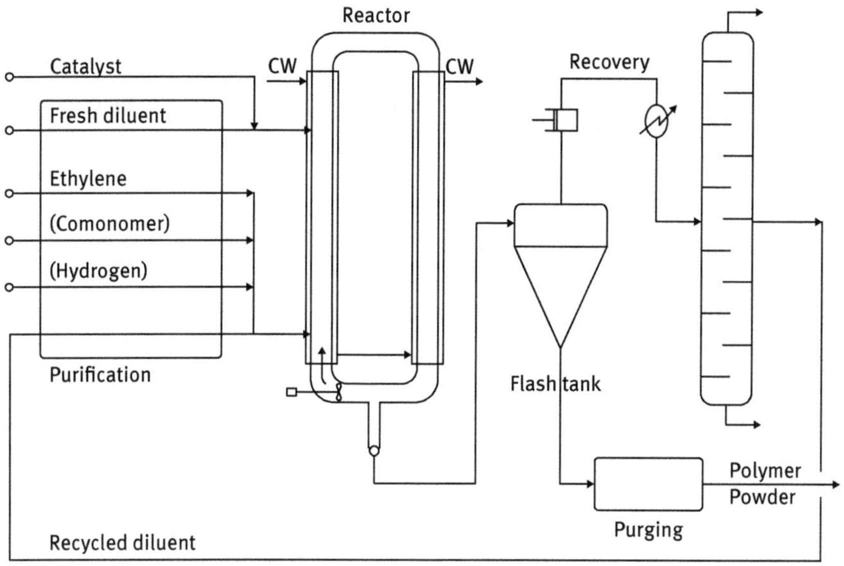

Fig. 1.1. Simplified flow sheet of the HDPE slurry loop process.

of concentrated slurry are released from each settling leg to the flash tank at regular intervals. The thus effected thickening of the slurry reaching the flash tank is saving capacity and energy in the recovery section.

The flash tank is maintained at a much lower pressure than the loop reactor, so most of the (hydrocarbon) liquid is flashed off there, and leaves the flash tank by the top outlet, towards the recovery section. The polymer powder is collected at the bottom of the flash tank, and purged from hydrocarbons in a powder purging step. From there it is transferred to the extruder building and pelletised.

2 The Loop Reactor Circulation Pump Power

The first concern of a young operating team, running a new slurry loop reactor, is to keep that reactor running!
When they are given a safe set of reactor conditions, and a known catalyst is fed at a moderate rate, they can maintain an accordingly moderate production rate, without much knowledge of reaction kinetics or catalyst behaviour. It is much more important to them to master the hydrodynamics of the reactor. The slurry therein is circulated at a high speed through the loop by a single axial pump. If that pump fails, the slurry will lose its velocity very fast and, within a few minutes' time, the polymer powder will settle in the bottom of the reactor as pillar-shaped lumps. There is no other way out then than to open the reactor and clean it, which operation takes at least a week, and is expensive and hazardous.
This trouble can be avoided by a good understanding of the reactor circulation pump behaviour and careful monitoring of the reactor pump power consumption.

2.1 Reactor Pump Circulating Pure Liquid

When starting up the reactor, before reagents are fed, the reactor pump will circulate a pure diluent liquid for some time, to allow to establish the required reactor temperature. The pump power consumed in these circumstances, at a certain temperature, is an interesting data, to be kept as a reference to judge the pump's future behaviour.

In general, the power consumed by a pump is the energy per unit time required to transfer a liquid from a point 1 to a point 2. It is given by Bernoulli's equation[1]. Four different resistances to flow between points 1 and 2 must be dealt with, viz. a difference in height, a difference in pressure, a difference in velocity, and the friction resistances from the pipe wall and in valves and fittings along the trajectory. When expressing these resistances as an energy to be supplied per unit mass of the pumped liquid, Bernoulli's equation looks like,

$$W = g(z_2 - z_1) + \frac{p_2 - p_1}{\rho_f} + \frac{v_2^2 - v_1^2}{2} + W_F\,,$$

where W is the total energy to be supplied to the liquid (J, kg^{-1}); z is the height of a point above an arbitrary horizontal reference plane (m); g is the acceleration by gravity (m, s^{-2}); p is the liquid pressure at a point (Pa); ρ_f is the density of the liquid (kg, m^{-3}); v is the velocity of the liquid at a point (m, s^{-1}), and W_F is the energy required to over-

1 Daniel Bernoulli, *Hydrodynamica*, 1738. Daniel Bernoulli (1700–1782) considered ideal fluids; no friction term appeared in his original equation.

come the fluid friction resistances in the pipe. It is supposed that the fluid properties are constant and that the velocity is uniform throughout a cross-section of the pipe.

Bernoulli's equation can be represented in different units. Pump characteristics given by the pump manufacturer usually have been measured with water as a liquid, and the energy supplied by the pump to the liquid is given as a *head*, i.e. an energy expressed in meters, independent of the density of the liquid. An energy thus expressed can be interpreted as the height at which one kilogramme of the fluid would own a static energy that is equivalent to the total energy W. The equation is then like,

$$H = (z_2 - z_1) + \frac{p_2 - p_1}{\rho_f g} + \frac{v_2^2 - v_1^2}{2g} + H_F ,$$

where H is the *total dynamic head* (m) and H_F is the *friction head.*

The equation can also be expressed in terms of pressure. The extra pressure to be supplied by the pump to the liquid to overcome the four resistances, is then,

$$\Delta p = \rho_f g(z_2 - z_1) + (p_2 - p_1) + \frac{\rho_f(v_2^2 - v_1^2)}{2} + F , \tag{2.1}$$

where Δp is the total extra pressure to be delivered by the pump to the liquid (Pa), and F is that part of the extra pressure due to the friction resistances.
It can be verified from a comparison of the units Pa and J, m^{-3} that the pressure difference Δp is equivalent to the energy to be delivered by the pump to the liquid per unit volume of liquid, or to the power to be delivered per unit of the volumetric liquid flow Q, (m^3, s^{-1}).

In the particular case of the loop reactor pump, the pipe, through which the liquid is pumped, forms a closed loop. The points 1 and 2 can be put arbitrarily somewhere along the loop, but must necessarily coincide to cover the whole path, Figure 2.1. Equation (2.1) is thus greatly simplified, the only resistance term left being the friction term,

$$\Delta p = F . \tag{2.2}$$

Δp is really the difference in pressure that can be measured in the liquid at the discharge and suction flanges of the reactor pump. Internal energy losses inside the pump should be taken into account by a pump efficiency η_p, which is typically around 75%.

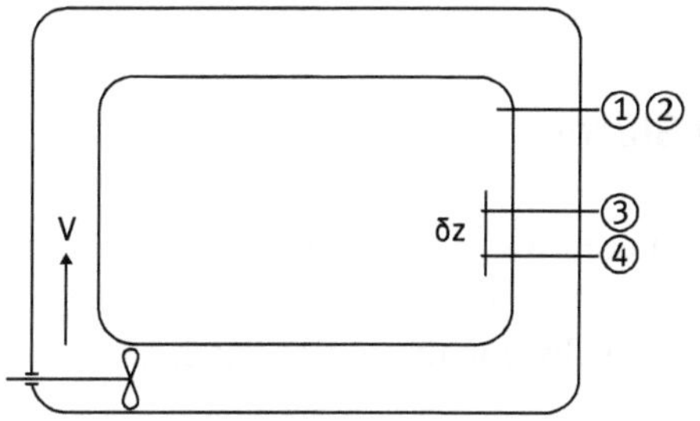

Fig. 2.1. Bernoulli's equation for the loop reactor pump.

Thereto should be added the energy losses in the gear box and mechanical transmissions, as also in the electric motor. The overall efficiency of the pumping group η will thus still be somewhat less.
The pump power as monitored by the reactor operator, and referred to in this chapter whenever *pump power* is mentioned, is the electric power consumption measured at the electric feed of the drive motor, and is thus equal to,

$$J = \frac{1}{\eta} Q\Delta p = \frac{1}{\eta} QF \,, \tag{2.3}$$

where J is the *reactor circulation pump power consumption*, (W).

The velocity at which the liquid or slurry is pumped through the loop is extremely high, of the order of magnitude of $10\,m,\,s^{-1}$, to obtain a good heat transfer at the inner reactor wall. So the flow is outspokenly turbulent, with Reynolds numbers between $4 \cdot 10^6$ and $5 \cdot 10^6$. The friction resistance F is mainly caused by the roughness of the pipe wall.
Wall friction in round pipes is well known from the literature as,

$$F = \rho_{\mathrm{f}} f \frac{L}{D} \frac{v^2}{2} \,, \tag{2.4}$$

where L is the length of the loop and D is the internal diameter of the loop pipe.
f is a *friction coefficient*, which, in turbulent flow, increases very strongly with the *roughness* ϵ of the internal pipe wall. Experimental values of f are known as a function of the relative roughness of the pipe wall ϵ/D, from Moody's diagram or from explicited empirical expressions.

The roughness of the loop reactor pipe walls is extremely important. Even if there is only one term in equation (2.2), the wall friction term, it is a large quantity and the pump power is considerable. New loop reactor pipes are polished and own very low roughnesses. If the roughness of the reactor wall is increased in operation by adherence of polymer particles to the wall, the reactor pump power will be strongly affected. A shift in roughness from $\epsilon = 0.01\,mm$ to $\epsilon = 0.10\,mm$, for example, would make the pump power increase by one third! Differences in pump power of about 10%, at apparently identical reactor conditions, are not exceptional. Polishing a commercial reactor in operation is hardly feasible, but, fortunately, the circulation of the slurry along the walls has some cleaning effect and keeps the pump power within acceptable limits.

2.2 Reactor Pump Circulating the Slurry

When in production, the reactor pump is circulating a slurry made up of reactor liquid and polymer particles. To determine the pump power is a matter of determining the pressure drop due to friction suffered by a slurry flowing through a length of pipe L.

This question is quite intricate, but has been described very clearly by Molerus and Wellmann in 1981[2]. Their results can be applied to the slurry in the HDPE loop reactor.

According to this work, the pressure drop can be represented as the sum of two pressure drops, one, F, due to the flow of the liquid, and a second one, Δp_{solids}, due to the presence of the solid particles.

Wall friction in the pipe occurs in a thin boundary layer along the wall where the velocity is very small. Solid particles can be assumed as not entering this layer, because the smaller particles are carried off by the main stream of fluid, away from the pipe wall, and the larger particles are too large to enter it. Consequently, the *friction losses at the wall*, F, when pumping a slurry, are caused by the liquid only, and equal to those found for the case of the pumping of a pure liquid, equation (2.4),

$$F = \rho_{\text{f}} f \frac{L}{D} \frac{v^2}{2}, \tag{2.5}$$

where v is now the *velocity of the slurry*, (m, s^{-1}).

The presence of the solid particles is causing an additional pressure drop, because they tend to lose their velocity by sedimentation and by collisions with other particles. They regain their velocity by the drag forces exerted by the liquid on the solid particles, but this action demands a certain energy from the liquid, which must be constantly replenished. Hence this other pressure drop, which is a fraction of the kinetic energy of the liquid, $\rho_{\text{f}} v^2/2$, that is consumed in the work exerted on the solid particles.

The *pressure drop due to the solid particles*, Δp_{solids}, can be represented quite simply for three different cases, according to the *volumetric solids concentration of the slurry*, C_0, that is the fraction of the slurry volume that is occupied by the solid matter.

For volumetric solids concentrations below around 25 *vol%*, $C_0 \leq 0.25$, the pressure drop is proportional to the solids concentration C_0,

$$\Delta p_{\text{solids}} = C_0 k_1 L \frac{\rho_{\text{f}} v^2}{2}. \tag{2.6}$$

For volumetric solids concentrations above around 25 *vol%* and below around 40 *vol%*, $0.25 \leq C_0 \leq 0.40$, the pressure drop contains two terms, one proportional to C_0, and a second one quadratic in C_0, because the particles are at closer distances to each other and the hinder is enhanced,

$$\Delta p_{\text{solids}} = C_0 k_2 L \frac{\rho_{\text{f}} v^2}{2} + C_0^2 k_3 L \frac{\rho_{\text{f}} v^2}{2}. \tag{2.7}$$

At solids concentrations above around 40 *vol%*, the slurry has become so dense that it cannot be pumped any more! The pressure drop tends to infinity.

The expressions (2.6) and (2.7) are empirical. The coefficients $k_1, k_2, k_3, (m^{-1})$, are not

2 Molerus O., Wellmann P., *A new concept for calculation of pressure drop with hydraulic transport of solids in horizontal pipes*, Chem. Eng. Sci., volume 36 (10), pp. 1623–1632, 1981.

dimensionless, and depend on the nature of the slurry (the liquid, the solid particles), *and also* on the diameter of the loop pipe D and the slurry velocity v. Molerus and Wellmann predict the value of these coefficients. Here, they are considered as coefficients that are obtained empirically from pump power data for each specific case. It should be noted that the coefficients are different from case to case, and, since they depend on D and v, are even different between reactors where identical reactor conditions are maintained, but of a different reactor geometry.

An example of a pump power curve as a function of the solids concentration is shown in Figure 2.2. The plot shows two pump power curves, at identical reactor conditions, shortly before and after a fouling of the reactor wall. The higher curve shows the increased pump power due to the increased roughness of the fouled wall. The pump power at zero % solids is the power consumed when only reactor liquid is circulated at the reactor temperature.

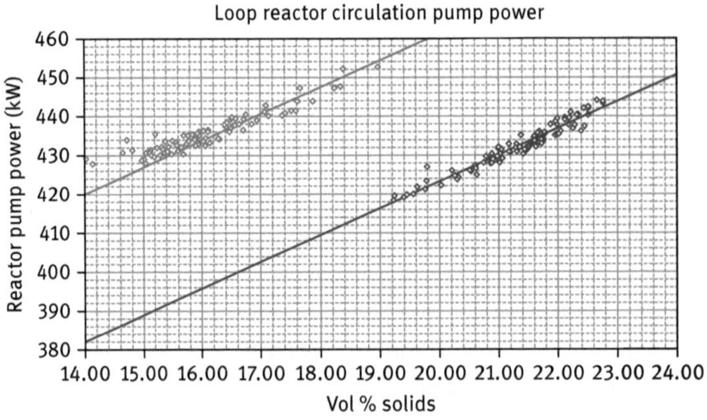

Fig. 2.2. Pump power versus vol% solids in an HDPE loop reactor. The higher curve shows the increased pump power, at equal reactor conditions, due to fouling of the reactor wall.

Experience shows that in HDPE slurry loop reactors the solids concentration should not be allowed to come near 25 *vol%*. The pump power increase is there so formidable and fast that equation (2.7) can never be verified, and pump failure is inevitable. So, for safe operation, C_O should be below 0.25, and the pump power is equal to, equations (2.5) and (2.6),

$$J = \frac{1}{\eta}Q(F + \Delta p_{\text{solids}}) = \frac{1}{\eta}Q\left(f\frac{L}{D}\frac{\rho_f v^2}{2} + C_O k_1 L\frac{\rho_f v^2}{2}\right). \tag{2.8}$$

However, the flow regime for solids concentrations $0.25 \le C_O \le 0.40$, described by equation (2.7), is visible in loop reactors where polypropylene is produced, in a slurry consisting of propylene liquid and polypropylene solids, so called bulk polymerisa-

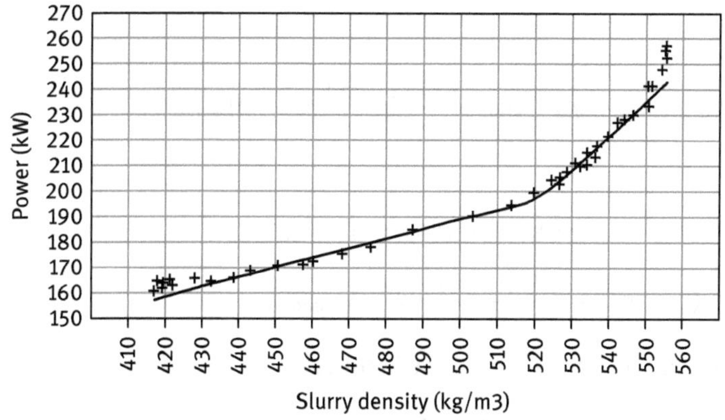

Fig. 2.3. Pump power as a function of solids concentrations in a loop reactor where polypropylene is produced. In the zone of concentrations $0.25 \leq C_0 \leq 0.40$ a more than linear increase of the pump power is visible.

tion. There, a stable operation of the reactor is possible in that zone of concentrations, as is shown in Figure 2.3.

A curve of the pump power can also be plotted versus the slurry density, which is a directly measured quantity, instead of versus the volumetric solids concentration, since there is a linear relation between the two,

$$C_0 = \frac{\rho_s - \rho_f}{\rho_p - \rho_f}, \tag{2.9}$$

where ρ_s is the *slurry density* and ρ_p the *polymer density*, but the solids concentration is more fundamentally linked to the phenomenon.

2.3 Upsets in Reactor Pump Power

2.3.1 A Realistic Reactor Pump Power Curve

In the previous section it was assumed that the composition of the slurry is absolutely uniform along the loop. The value of the pump power should consequently be constant in time. In reality, differences in composition will affect the pump power, so that, as seen from a fast recorder, it is not constant but shows fast fluctuations of a certain amplitude. This *pump power amplitude* is an important feature to follow during the operation of the slurry loop reactor.

A great part of those fluctuations can be explained by differences in the density of the slurry along the loop. If in a small zone $\Delta z = z_4 - z_3$ in the loop, Figure 2.1, the density of the slurry is equal to $\rho_s + \Delta\rho_s$ instead of ρ_s, then Bernoulli's equation (2.1), written for the slurry,

$$\Delta p = \rho_s g(z_2 - z_1) + (p_2 - p_1) + \frac{\rho_s\left(v_2^2 - v_1^2\right)}{2} + F + \Delta p_{\text{solids}} ,$$

should read,

$$\begin{aligned}\Delta p = \rho_s g(z_2 - z_4) + (\rho_s + \Delta\rho_s)g(z_4 - z_3) + \rho_s g(z_3 - z_1) \\ + (p_2 - p_1) + \frac{\rho_s\left(v_2^2 - v_1^2\right)}{2} + F + \Delta p_{\text{solids}} ,\end{aligned}$$

or, taking into account that the points 1 and 2 are identical,

$$\Delta p = \Delta\rho_s g(z_4 - z_3) + F + \Delta p_{\text{solids}} .$$

Two cases should be discerned. If the zone $\Delta z = z_4 - z_3$ is located in a reactor leg where the slurry is flowing upward, then $\Delta z = z_4 - z_3 > 0$. If the slurry is flowing downward in that zone, then $\Delta z = z_4 - z_3 < 0$. The pump power thus owns two values, depending on whether the zone Δz is flowing upward or downward,

$$\Delta p = \pm\Delta\rho_s g|\Delta z| + F + \Delta p_{\text{solids}} ,$$

and, since the slurry is circulated at a constant speed around the loop, the value of Δp will periodically adapt the one value or the other, at the frequency of the circulation. Since many more differences in slurry density occur along the loop, the pump power curve will show an intricate fluctuating pattern, hence the existence of a pump power amplitude. A fast recording of the reactor pump power allows to discern the periods in that pattern, and thus the exact frequency or speed at which the slurry is circulated.

2.3.2 Reactor Fouling

Reactor fouling is one of the constant threats to the safe operation of a slurry loop reactor.

The original meaning of the term *fouling* is related to the fouling of the reactor wall by deposits from the reactor liquid, like all heat exchanger walls are subject to fouling. The fact of fouling is announced by the pump power amplitude's growing wider, before the pump power will increase in average value. This should be noticed by the operator, and interpreted as a warning for some upset to come.

Also other sorts of upsets are announced by an increased pump power amplitude, and so the meaning of the term fouling has become broader. It is now indicating all upsets that are noticed by the pump power amplitude.

Fouling of the reactor wall is the genuine type of fouling. Polyethylenes own a broad molecular weight distribution (MWD), starting from molecular weights below 1,000 grammes per mole. So part of the polymer, on the low molecular weight (LMW) side, will be dissolved in the reactor liquid, at a concentration according to the temperature of the bulk of the slurry. The reactor walls are necessarily kept at a lower temperature to remove the heat of reaction, and it is possible that LMW polymers, or oligomers, are deposited at those walls. This phenomenon is not so dramatic as to cause serious reactor fouling at every temperature dip that may happen, but it should be paid attention to. Isobutane is often selected as the diluent exactly because it dissolves less oligomers. The heavier the diluent is, the greater the chance of fouling. The use of comonomers, dissolved in the diluent, to produce polyethylenes of lower densities, makes the reactor liquid heavier, and more prone to fouling. It is indeed one of the great limitations to the slurry loop process that a range of lower polymer densities cannot be produced because the required high comonomer concentrations would induce fouling. Several experimental *fouling curves* are known, delimiting zones of safe operation and fouling in a diagram of polymer density versus reactor temperature, Figure 2.4. Polymer density is here an indirect variable, it is rather the comonomer concentration that is determining the fouling curve at a given temperature. There is not one universal fouling curve. The curves depend on the catalyst type, because different catalyst families produce different MWD's. Chromium oxide catalysts are producing the broadest MWD, and will impose the strongest limitations. A metallocene catalyst, on the other hand, can produce Low Density materials in a loop[3]. The fouling curve will also depend on the average molecular weight of the polymer, or melt index. All this explains why Figure 2.4 is not showing a neat curve, but a grey zone where most fouling curves will be found. And operating above the fouling curve is not an absolute safeguard for all possible fouling. E.g., a typical operation error, which will be dearly paid for, is the use of a chromium catalyst in the presence of triethylaluminium (TEAl), the Ziegler–Natta cocatalyst. The combination of the two will produce such quantities of oligomers as to induce reactor fouling very soon.

A second cause of an increased pump power and/or -amplitude is a too high level of *static electricity* present in the slurry. The polymer particles will attract each other, and the slurry behaviour will show a much bigger resistance to flow. Static electricity is always generated by friction when a liquid of very low electric conductivity, like the reactor diluent, is pumped at high speeds. The solution to this problem is to dose an additive that will enhance the conductivity of the diluent, and thus allow static charges to be drained via the reactor walls. Those additives stem from the area of aircraft fuels that are pumped at high speed to fill the aircraft fuel tanks.

A similar effect is noticed when *the polymer particles are too fine*. Very fine polymer particles will attract each other by static charges and/or Van der Waals forces. The fineness of the polymer is mainly determined by the catalyst's particle size distribu-

3 Whether a catalyst will involve a high risk of fouling at a certain polymer density also depends on its ability to incorporate the comonomer. This is treated in Chapter 9, where also is shown how the reactor conditions can be optimized for the production of lower densities at the safe side of the fouling curve.

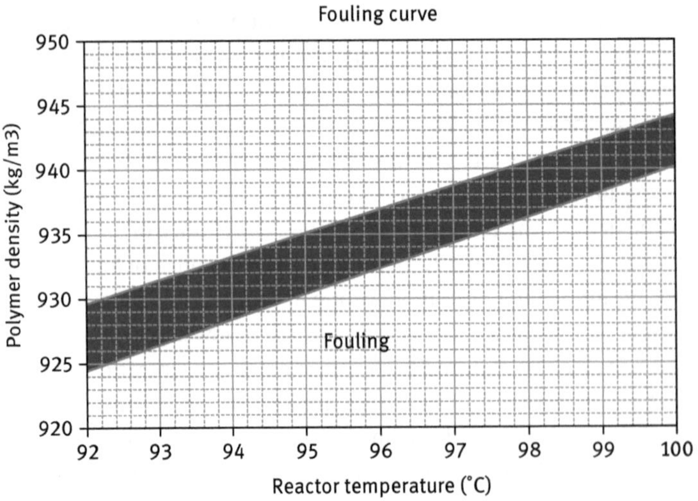

Fig. 2.4. Experimental fouling curve, delimiting zones of safe operation and fouling in a diagram of polymer density versus reactor temperature. The fouling curve is not shown as a neat curve, but as a grey zone where most fouling curves will be found.

tion. A typical feature by which this phenomenon can be recognized is that the reactor pump power will jump to a higher level very suddenly, without any ramping, and will later at once drop back to lower values.

Fouling of the worst kind is met *when the bubble point of the reactor liquid is exceeded*, either by a too low reactor pressure, or a too high temperature, at a given concentration of the light components. Components contributing to a lower bubble pressure are ethylene, hydrogen, and ethane if not properly purged from the system. The gas bubbles generated in the liquid when the bubble point is exceeded, will contain higher concentrations of these light components. The polymerisation reaction will be out of control, and the polymer will show the weirdest morphology, no particles, but ropes and strings, and whatever names can be given to the shape of the polymer. Those will plug the reactor outlet and its downstream equipment. It follows that the control of a stable ethylene concentration is one of the greatest conditions to the process-safe operation of a loop reactor.

2.3.3 Hot Spots

A last upset, typical for a loop reactor, and related to the pump's circulating of the slurry, is caused by the so-called hot spots. A hot spot is, in general, an area in a chemical reactor where, due to a lack of mixing, unwantedly high temperatures occur. They are a risk in polyethylene loop reactors where, as is often the case, a discontinuous catalyst feed system is used, injecting small quantities of catalyst at regular intervals.

Due to the highly turbulent flow of the slurry, there will be no, or very little, mixing along the length of the loop. An accumulation of catalyst will occur at a point in the slurry where the discontinuous catalyst feed is injected in phase with the period at which the slurry is circulated, and thus repeatedly at the same spot. Such a *hot spot* will show a higher catalyst and polymer concentration, and a higher temperature. Hot spots affect the catalyst activity badly because the local high concentrations of catalyst will exhaust the local ethylene concentrations, whilst the lack of catalyst in other areas will lower the reactor's efficiency[4].

A method to avoid hot spots is presented here, but can be saved for later reading.

2.3.3.1 Conditions at which a Hot Spot is Created

If T_S is the *period at which the slurry is circulated*, and L is the length of the reactor loop, then the velocity of the slurry, v, is equal to $v = L/T_S$.

A *one-dimensional coordinate system* is defined, consisting of one x-axis, coinciding with the loop, of which the origin is moving along the loop at the (constant) speed of the slurry $v = L/T_S$, the positive sense of the x-axis pointing in the direction of the slurry velocity. In other words, the loop reactor is seen as a one-dimensional system, and the observer is traveling through the loop together with one point of the slurry, just like a passenger on a train is observing the landscape along the railroad.

The point where the catalyst is injected is moving at a negative speed equal to $-L/T_S$. Let T_C be the *period of the discontinuous catalyst injection.*
Consider an injection at time $t = 0$ at point $x = 0$. The second injection at time $t = T_C$ enters at point,

$$x = -\frac{L}{T_S}T_C .$$

The third injection at time $t = 2T_C$ enters at point,

$$x = -\frac{L}{T_S}2T_C .$$

The $(n + 1)$th injection at time $t = nT_C$ will enter at point,

$$x = -\frac{L}{T_S}nT_C .$$

A hot spot will be created when the $(k + 1)$th injection, after the first injection at $t = 0$ and $x = 0$, that is k injections later, enters the slurry at the same point $x = 0 - mL = -mL$, where m is a positive integer. The reactor, indeed, forms a closed loop

4 An example of the effect of hot spots on the catalyst activity is found in Section 8.4, Figure 8.5. There the catalyst productivity (to be defined later) at catalyst residence times around 1.0 and 1.1 h is very much below expectation because the frequency of the catalyst injections is in phase with the frequency at which the slurry is running the loop, creating a number of hot spots.

of length L, and all points $x = -mL$ coincide.
This will be the case if the injection period T_C fulfils the condition,

$$-mL = -\frac{L}{T_S} kT_C \,,$$

or,

$$\frac{T_C}{T_S} = \frac{m}{k} \,. \tag{2.10}$$

The condition for the generation of a number of hot spots is that the ratio T_C/T_S is equal to the ratio m/k of two positive integers m and k. This condition is independent of the time of injection. If it holds for a first injection at time $t = 0$, it will hold for later injections at times $t = kT_C$, $t = 2kT_C$ etc. A periodic injection of catalyst will occur at the hot spot. The period will be kT_C.
Other ratios jm/jk where jm and jk are the same multiple of m and k do not represent a new hotspot, because these injections have already been included in the group of injections at the ratio m/k.

The conclusion is that, in order to avoid hot spots, periods of catalyst injection T_C where the ratio T_C/T_S is equal to a ratio m/k of two positive integers m and k, must be avoided.

2.3.3.2 Interpretation of the Integers *m* and *k*

The injection period at a hotspot in the slurry, with $T_C/T_S = m/k$, is equal to

$$kT_C = k\frac{m}{k} T_S = mT_S \,.$$

Each time after m circulation periods, another injection of the same quantity of catalyst will enter the slurry at the hot spot. The integer m may be called the *order of the hot spot*. The smaller the value of m, the larger the accumulation of catalyst will be.

The condition (2.10), $T_C/T_S = m/k$, is independent of the time of injection. If it holds at time $t = 0$, it also holds at time $t = T_C$. There it will start a new periodic injection with a period kT_C, and injection times $T_C, (1 + k)T_C, (1 + 2k)T_C, \ldots$ etc, i.e. another hotspot.
More hotspots will thus be created:
from time $t = 2T_C$ with injection times $2T_C, (2 + k)T_C, (2 + 2k)T_C, \ldots$ etc;
from time $t = 3T_C$ with injection times $3T_C, (3 + k)T_C, (3 + 2k)T_C, \ldots$ etc;
and finally,
from time $t = (k - 1)T_C$ with injection times $(k - 1)T_C, (2k - 1)T_C, (3k - 1)T_C, \ldots$
The condition (2.10), fulfilled for one value of m/k thus results in k hot spots of order m.
The integer k may be called the *number of hotspots*.

The distance between two adjacent hotspots that thus have been created at one value of m/k is equal to,

$$vT_C = \frac{L}{T_S} T_C = \frac{mL}{k} \,.$$

So the distances from the first hotspot created from time $t = 0$ to the next $(k-1)$ hotspots will be,

$$\frac{mL}{k}, \frac{2mL}{k}, \ldots, \frac{nmL}{k}, \ldots, \frac{(k-1)mL}{k},$$

with $n = 1, 2, 3, \ldots, (k-1)$.
These distances are $(k-1)$ different multiples of L/k. So the positions of the k hotspots will be nicely distributed along the length of the slurry at equal distances L/k.

All this could be summarized by defining a *Hot Spot* m/k as the fact that T_C/T_S is equal to the ratio m/k, and gives rise to k hot spots located at k equal distances L/k along the loop, where catalyst is injected at a period equal to m circulation periods T_S. The worst hot spot is hot spot 1/1 created when $T_C = T_S$, and all catalyst is injected at one single point in the loop.

2.3.3.3 Margins to be Respected around m/k Values

In theory, all possible m/k values are to be avoided for the ratio T_C/T_S to prevent the creation of hotspots. In practice, the periods T_C and T_S are subject to variations or errors, and, since the speed of the slurry is extremely high, small deviations in the timing of the catalyst injection entrain large variations in distance between the injection points in the slurry. So, for practical purposes, a certain distance δ should be respected between the calculated value of T_C/T_S and any ratio m/k, in order to keep a safe distance between the injection points in the slurry.
If,

$$\frac{T_C}{T_S} = \frac{m}{k} \pm \delta,$$

and a first injection at time $t = 0$ enters at point $x = 0$, the next injection near a possible hotspot m/k at time $t = kT_C$ will enter at point,

$$x = -\frac{L}{T_S} kT_C,$$

or,

$$x = -Lk\left(\frac{m}{k} \pm \delta\right).$$

The distance between the injection point at time $t = kT_C$ and the to be avoided hotspot will thus be,

$$\Delta x = \mp Lk\delta.$$

The operating engineer can decide what distance Δx to the avoided hotspot should be respected, and then δ should be equal to,

$$|\delta| = \frac{|\Delta x|}{kL}. \tag{2.11}$$

The safe distance Δx can also be expressed as a distance in time Δt, with reference to the precision at which an injection period T_C can be timed by the process control system,

$$|\Delta t| = \frac{|\Delta x|}{v} = \frac{|\Delta x|}{L/T_S} ,$$

or,

$$|\delta| = \frac{|\Delta t|}{kT_S} . \tag{2.12}$$

The required value of the safety distance δ is a function of k. It is smaller for larger values of k. To maintain the same distance in meters or seconds from the avoided hot-spot at different k values, the according δ values should be selected k times smaller than the δ_1 value selected at $k = 1$. An equal safety margin will thus be attained in all cases.

In fact, hot spots of larger and larger values of k, are less and less harmful. A value of $k \to \infty$ even represents the perfectly even spread of the catalyst along the loop!

An alternative criterion to determine the safety margin δ follows from the consideration that δ should be large enough lest the injection point should move too slowly along the length of the slurry, and form a kind of creeping hotspot.

In other words, the injection point, which is moving by a step Δx after every injection period kT_C should have covered the whole slurry length L within a maximum time t_{max}.

The number of steps taken within time t_{max} is equal to t_{max}/kT_C.

The distance covered should be at least equal to L,

$$\frac{t_{max}}{kT_C} \Delta x \geq L ,$$

or, equation (2.11),

$$\frac{t_{max}}{kT_C} Lk\delta \geq L ,$$

and,

$$\frac{t_{max}}{T_C} \delta \geq 1 .$$

So the minimum value required for δ is,

$$\delta \geq \frac{T_C}{t_{max}} = \frac{T_C}{T_S} \frac{T_S}{t_{max}} = \left(\frac{m}{k} \pm \delta \right) \frac{T_S}{t_{max}} ,$$

or,

$$\delta \geq \frac{m}{k} \frac{T_S}{t_{max}} \frac{1}{1 \mp \frac{T_S}{t_{max}}}$$

Since t_{max} can be thought to be of the order of magnitude of the catalyst residence time, it can be assumed that $T_S \ll t_{max}$ and the requirement for δ is then that,

$$\delta \geq \frac{m}{k} \frac{T_S}{t_{max}} . \tag{2.13}$$

At a certain value of δ, the length of the slurry L will be covered in a time t_L, equal to,

$$t_L = \frac{L}{\Delta x} kT_C = \frac{T_C}{\delta} = \frac{m}{k} \frac{T_S}{\delta} .$$

Notations

C_0	volumetric concentration of solids in the reactor, (–).
D	internal diameter of the loop pipe, (m).
f	friction coefficient, (–).
F	extra pressure to be delivered to the liquid to overcome the fluid friction resistances when pumping it from a point 1 to a point 2, (Pa).
g	acceleration by gravity, (m, s^{-2}).
H	total dynamic head: total energy to be supplied per kg of liquid to pump it from a point 1 to a point 2 expressed as a head, (m).
H_F	friction head: energy to be supplied per kg of liquid to overcome the fluid friction resistances when pumping it from a point 1 to a point 2, expressed as a head, (m).
j	a positive integer, (–).
J	reactor circulation pump power consumption, measured at the electric feed of the reactor pump drive motor, (W).
k	a positive integer, number of hot spots, (–).
k_1, k_2, k_3	coefficients, (m^{-1}).
L	length of the loop pipe, (m).
m	a positive integer, order of a hot spot, (–).
n	a positive integer, (–).
p	liquid pressure, (Pa).
Q	volumetric liquid flow, (m^3, s^{-1}).
t	time, (s).
t_{max}	maximum time within which the catalyst injection point, moving along the slurry, should have covered the whole slurry length L, (s).
t_L	time within which the catalyst injection point, moving along the slurry, is covering the whole slurry length L, (s).
T_c	period of a discontinuous catalyst injection in the loop, (s).
T_s	period at which the slurry is circulated around the loop, (s).
v	velocity of the liquid, or slurry, (m, s^{-1}).
W	total energy to be supplied per kg of liquid to pump it from a point 1 to a point 2, (J, kg^{-1}).
W_F	energy to be supplied per kg of liquid to overcome the fluid friction resistances when pumping it from a point 1 to a point 2, (J, kg^{-1}).
x	a one-dimensional coordinate along the length of the loop, of which the origin is moving along the loop at the (constant) speed of the slurry $v = L/T_s$, the positive sense of the x-axis pointing in the direction of the slurry velocity, (m).
z	height of a point above an arbitrary horizontal reference plane, (m).
δ	a distance to be respected between the calculated value of T_c/T_s and the ratio m/k, to avoid hot spots, (–).

Δp	total extra pressure to be delivered to a liquid to pump it from a point 1 to a point 2, (Pa).
Δp_{solids}	fraction of the pressure drop to be deliverd to a slurry to pump it from a point 1 to a point 2, due to the presence of the solid particles, (Pa).
Δt	a distance to be respected in time between the time of injection and an injection period T_c that would create hot spots, (s).
Δx	a spatial distance to be respected between the injection point of the catalyst in the slurry and a possible hot spot, (m).
Δz	a small zone in the loop where the density of the slurry is different from the density in the rest of the loop, (m).
$\Delta \rho_s$	a small difference in the density of the slurry along the loop, (kg, m^{-3}).
ϵ	roughness of the internal pipe wall, (m).
η	overall efficiency of the pumping group, taking into account energy losses inside the pump, in the gear box and mechanical transmissions, and in the electric motor, (–).
η_p	pump efficiency taking into account internal energy losses inside the pump, (–).
ρ_f	density of the liquid in the reactor, (kg, m^{-3}).
ρ_p	density of the polymer in the reactor, (kg, m^{-3}).
ρ_s	density of the slurry in the reactor, (kg, m^{-3}).

3 The Functioning of the Settling Legs

The settling legs are a vital feature of most of the slurry loop reactors where ethylene is polymerised. They are simply a number of vertical tubes connected to one of the lower elbows of the loop, pointing downward and fully open to the loop itself. The polymer particles will tend to settle into these tubes, where they will concentrate in the bottom. This bottom is connected to the flash tank, and is shut off by a valve that is normally closed, but periodically opened for just a few seconds. The concentrated slurry is thus intermittently released to the flash tank, where reactor liquid and polymer are separated by the liquid's flashing off.

Settling legs were patented by J. S. Scoggin in 1966[1]. The first advantage brought by the settling legs is that they concentrate the slurry from a maximum of around 40 *wt*% of polyethylene in isobutane to over 50 *wt*%, which means that the size of the diluent recovery section and the flow rate of recycled diluent through the plant are significantly reduced.

Many attempts were made to replace the settling leg system, e.g. by a continuous take-off valve and especially by placing this take-off point at the outer side of a loop reactor elbow, where the slurry is believed to be concentrated by inertial forces. Or by the use of hydrocyclones to concentrate the slurry with. It will, however, be difficult to design a more efficient system than the settling legs because they handle the slurry at concentrations where it can hardly be pumped any more, and they provide a simple intrinsic system to control the pressure, the solids concentration and the residence time in the reactor.

The capacity of the settling legs, i.e. the rate of solids they can remove from the reactor, is important to the production capacity of the reactor, because in steady operation, the rate of solids removed by the legs must equal the rate of solids produced. The settling legs' capacity can be the limiting factor to the reactor capacity, especially when the polymer particles are not settling well enough. It is often thought, and so did the inventor, that the capacity of the legs is proportional to their number or to the total of their cross sectional areas. It will soon become clear that this is not altogether true, because settling is only one of the two mechanisms by which the solids are removed via the legs.

In this chapter is presented how the settling legs are functioning and how they affect the operation of the loop reactor. This knowledge is vital for those who are operating slurry loop polymerisation reactors equipped with settling legs, since one of the first conditions to operate a slurry loop reactor is to manage a stable control of the solids concentration. A more profound discussion of the settling of the polymer in the settling legs is given in the next Chapter 4.

1 US Patent 3,242,150, *Method and Apparatus for the recovery of solid olefin polymer from a continuous path reaction zone*, by Jack S. Scoggin, filed March 1960, patented March 1966.

3.1 The Functioning of the Settling Legs

3.1.1 Description and Definitions

A number of simplifying assumptions are made to describe the functioning of the settling legs without affecting the practical usefulness of the model.

1. Steady state conditions are assumed. The discontinuity created by the periodical opening of the settling legs is happening at a time scale much smaller than the residence times in the reactor, so the operation can be regarded as continuous and in the steady state.
2. Production rate and reactor solids concentration are treated as independent variables, without asking yet how they can be achieved in the reactor.
3. The reactor slurry is treated as consisting of two phases, pure polymer and pure liquid, in which polymer density and reactor liquid density are considered as constants. This treatment is offering sufficient accuracy. The compressibility of the liquid is thus discarded. So is the compressibility of the polymer. Moreover, the supposed swelling of the polymer particle by absorption of diluent liquid, which would affect the density and the size of the polymer particle, is not considered. Even if this swelling is often mentioned in this context, no need was felt here to do so in order to represent practical data. Also, measurements of the polymer particle porosity, which would absorb the diluent, show that this porosity is negligible. The polymer density is of course influenced by the reactor temperature.

The functioning of the settling legs is first described in words, introducing a certain terminology, and illustrated in Figure 3.1.

The settling legs are steered by the reactor pressure. Since the reactor is completely filled with a slurry, and no gas phase is present, the reactor pressure will be equal to the hydraulic pressure of the liquid. This pressure is generated by the recycled diluent pump that is constantly feeding recycled diluent to the reactor. It is determined by the characteristic curve of the pump, and is rising constantly, as more liquid is continuously pumped into the constant volume of the reactor vessel.
The operating pressure of the reactor is a maximum pressure in the loop that is controlled by the opening of one settling leg every time it is reached. By opening of the leg, it is meant that the bottom valve or take-off valve is just opened for a few seconds. A limited volume of slurry will thus be released to the flash tank, and the pressure will drop at once by one or two bars. But it will rise again as the recycled diluent feed is never interrupted. The settling legs are used in sequence, so when the maximum pressure is reached again, the next leg will be opened. When all legs have had their turn, the first leg wil be opened again and so forth ... The result is that the controlled reactor pressure is showing a toothed curve in time, with peaks and instant drops every time the set point is reached and a leg is opened. These pressure fluctuations can be managed by the process because they are limited due to the compressibility of the liquid.

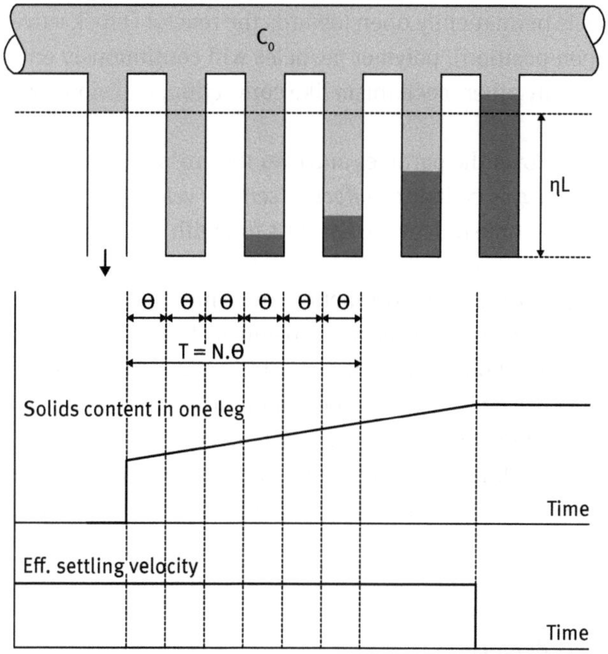

Fig. 3.1. The functioning of the settling legs.

The diluent is not an ideal liquid, and especially when ethylene is dissolved therein, it shows a considerable compressibility. (Which, however, as mentioned above, is not taken into account in this chapter's calculations.)

At the opening of a leg[2], a fraction η of the leg's volume is emptied towards the flash tank. This fraction η is called the *emptied fraction*.

To allow stable operation of the reactor, η should be large enough so that all settled polymer in the leg is taken off at each opening. If not, an accumulation of solids would occur, since at every opening another quantity of settled polymer would not be evacuated. The total solids concentration in the reactor would rise continuously, and, if not stopped, lead to a plugged reactor.

This implies that the solids concentration in the upper volume fraction $(1-\eta)$ of the leg is always assumed to be equal to the *volumetric solids concentration in the reactor*, C_0.

After the opening, the emptied volume is instantaneously filled again with reactor slurry, at the reactor solids concentration C_0. It is important to take note of this, because, besides settling, this is the second mechanism by which solids are removed from the reactor.

2 This opening is often called 'firing'. Here the use of military language in technology is avoided, as technology should be a civil activity aiming at general prosperity.

As a settling leg in service is permanently open towards the reactor (block valves between reactor and leg in open position), polymer particles will continuously enter the leg by gravity settling or by any other mechanism like convection, collisions with other particles, etc.
The net vertical velocity, averaged over the particles and over the cross sectional area, by which the particles enter the leg, is called the *effective settling velocity*, v_{sett}. It is, at first, supposed to be constant in time unless the leg runs full with settled polymer particles.
This saturation of the leg is, according to good practice, prevented by keeping the *cycle period*, T, i.e. the period between two consecutive openings of one and the same settling leg, small enough. So throughout this discussion, it is assumed that the legs never run full, and that v_{sett} is constant in time with no limitations.

The cycle period T equals N times the *opening period*, θ, i.e. the period between two consecutive openings of two settling legs in sequence, N being the *number of settling legs in service*. θ is the period that is noticed in the toothed curve of the reactor pressure versus time.

3.1.2 The Overall Reactor Mass Balance

In the steady state, the *sum of all feeds to the reactor*, ΣF_i, will leave the reactor by the settling legs either as polymer, or as reactor liquid. The polymer flow from the reactor is equal to the *production rate*, P. The liquid flow from the reactor is totally recycled, and equal to the *total recycled diluent flow*, D_{R},

$$\Sigma F_i = D_{\mathrm{R}} + P\,.$$

The *weight fraction of solids in the slurry discharged by the settling legs*, w, is,

$$w = \frac{P}{D_{\mathrm{R}} + P} = \frac{P}{\Sigma F_i}\,. \tag{3.1}$$

The average volumetric fraction of solids in the legs at the end of each cycle period, C_{T}, is easily obtained from w,

$$C_{\mathrm{T}} = \frac{P\rho_{\mathrm{f}}}{D_{\mathrm{R}}\rho_{\mathrm{p}} + P\rho_{\mathrm{f}}}\,, \tag{3.2}$$

where ρ_{p} and ρ_{f} are the densities of the polymer and the liquid respectively.
It should be noted that an overall mass balance of the reactor cannot distinguish between settling legs, and that the values w and C_{T} in equations (3.1) and (3.2) are necessarily average values over the total number of settling legs in service.
Also, to be more exact, the vulgar expression *in the legs* should be read as *in the total emptied volume of the legs in service*.
It is also important to note that equations based on steady state conditions are only

valid under such conditions. Often quantities derived from such equations are calculated online in process computers. It should be realised that those values are meaningless when no stable conditions prevail. This is too often forgotten when interpreting online data.

3.1.3 The Volumetric Output Equation

The volume of slurry discharged from a settling leg each time it is opened, can be stated in two ways: in terms of the settling leg geometry, or in terms of the polymer and diluent contents,

$$\eta \frac{\pi D^2}{4} L = \left(\frac{P}{\rho_p} + \frac{D_R}{\rho_f} \right) \theta \,,$$

where D is the *diameter of the settling leg*, and L is the *total length of the settling leg*. Hence follows the *volumetric output equation*,

$$\frac{\eta}{\theta} \frac{\pi D^2}{4} L = \frac{P}{\rho_p} + \frac{D_R}{\rho_f} \,. \tag{3.3}$$

The ratio η/θ is called the *opening frequency* and is characteristic for the operation of the settling legs. It is the number of times per unit time that the total settling leg volume $(\pi D^2/4)L$ is discharged to the flash tank. In all calculations concerning the volumetric output from the legs, the opening frequency appears, and never so the single components of the ratio.

Indeed, the emptied fraction η can be set between certain limits, 0.8 to 0.9 being a good value to have all concentrated slurry discharged, by controlling the speed at which the product take-off valves at the outlet of the legs are travelling, but the corresponding opening period θ is then determined by the opening frequency η/θ. For, if at each opening a smaller or greater volume of slurry is released according to a smaller or greater value of η, the consequent pressure drop in the reactor will be smaller or greater too. It will take less or more time to reach the set pressure again, and θ will be smaller or greater as well.

It is a common misunderstanding that the settling legs' capacity is proportional to their volume, and that the capacity is increased by making the legs longer at equal diameter. The emptied volume can indeed be greater from a larger leg, but after emptying it will take more time to reach the set pressure again, so the opening period θ will be greater too. The opening frequency is inversely proportional to the volume of the leg, and is imposed by the production and diluent feed rates.

This becomes more clear when, by use of equation (3.1), it is derived from equation (3.3) that the volume, discharged by the settling legs per unit of time, is determined by the production rate in the first place, since the solids concentration in the legs and the polymer and liquid densities show much less variability,

$$\frac{\eta}{\theta} \frac{\pi D^2}{4} L = P \left(\frac{1}{\rho_p} + \frac{1-w}{w \rho_f} \right) . \tag{3.4}$$

Consequently, the opening frequency of a given settling legs' system can hardly be controlled, but is proportional to the production rate. When starting up the loop reactor, and boosting the production rate to its full capacity, the opening frequency will increase because the diluent and monomer feeds to the reactor will be increased, and the set maximum pressure will be reached sooner after each opening of a leg.
It is this proportionality that is intrinsically present in the system, and that makes the settling legs a smooth system to control the reactor pressure and the solids concentration.

The volumetric output equation (3.3) is used in practice to calculate the emptied fraction η, online[3], or from historical data,

$$\eta = \left(\frac{P}{\rho_\mathrm{p}} + \frac{D_\mathrm{R}}{\rho_\mathrm{f}}\right)\left(\frac{\pi D^2}{4}L\right)^{-1}\theta\,. \tag{3.5}$$

Also here, stable conditions should prevail to obtain meaningful values.
The exact value of the settling leg volume, or, more specifically, of the settling leg length L, is not very critical. One can define the length L somewhat arbitrarily, e.g. including the top valves between the legs and the reactor loop, or not. Important is to maintain this value of the volume (length) throughout all calculations, and to use values of η that have been calculated consistently with the assumed leg volume.

3.1.4 The Solids Removal Rate Equation

The rate of solids removed from the reactor by entering one settling leg, can be quantified according to the description given in Section 3.1.1.

One mechanism by which the solids are removed from the reactor, is the displacement of concentrated slurry by fresh slurry from the reactor, entering the emptied leg volume at every opening of the leg. This mechanism is called *displacement*. It is the displacement mechanism that is often left out of consideration when is stated, wrongly, that the settling legs' capacity is proportional to its total cross-sectional area, or to the number of legs.

The solids also enter the leg by (effective) settling, as polymer particles pass the cross sectional area at the height ηL, at the constant average (effective settling) velocity v_sett. This second mechanism is called *sedimentation*. The rate of solids removed from the reactor through this mechanism is proportional to the total cross-sectional area.

The rate of solids removed by the settling legs, can now be stated as the sum of the contributions from both mechanisms. From each leg that is opened, the solids

---

3 The production rate P is calculated in the process computer from a mass balance over the reactor, or from a precise energy balance. The liquid density ρ_f is calculated from the measured liquid composition. The opening period θ is recorded in the process computer.

contained in the leg at the end of one cycle period T are released. Those solids have entered the leg by displacement at its previous opening, and by continuous sedimentation during the cycle period. One such opening happens after each opening period θ. Under stable conditions the rate of solids removed by the legs must equal the production rate during the opening period. Thus,

$$\rho_p \eta \frac{\pi D^2}{4} L C_0 + \rho_p \frac{\pi D^2}{4} C_0 v_{sett} T = P\theta .$$

Or, with $T = N\theta$,

$$\rho_p \frac{\eta}{\theta} \frac{\pi D^2}{4} L C_0 + \rho_p \frac{\pi D^2}{4} C_0 v_{sett} N = P . \tag{3.6}$$

This equation (3.6) or *solids removal rate equation*, is a mass balance for solid polymer. It describes how the production rate is taken off the reactor by displacement and sedimentation.

It shows how displacement depends on the opening frequency, and sedimentation on the number of legs in service.

It is a very important equation because it establishes the link, which is often missed, between the reactor solids concentration C_0 and the operation of the settling legs.

All variables in the equation are known or measured[4], except for the effective settling velocity v_{sett}. So equation (3.6) allows to obtain v_{sett} empirically from data sets extracted from stable operation periods,

$$v_{sett} = \frac{P - \frac{\eta}{\theta}\frac{\pi D^2}{4} L \rho_p C_0}{N \frac{\pi D^2}{4} \rho_p C_0} . \tag{3.7}$$

3.2 Correlations for the Effective Settling Velocity

Empirical values of the effective settling velocity can be calculated from plant data by the use of equation (3.7), and correlated to yield correlations that are used in practical models.

3.2.1 General Correlation

In one study, 93 data points were collected at stable conditions, covering different product grades with varying powder bulk densities, produced from Ziegler–Natta and chromium oxide catalysts at various production rates and with 3 to 6 settling legs in

4 The solids concentration is calculated from the measured slurry density, the liquid density, and the polymer density.

service. The diameter of the settling legs was 10 inch. The opening frequency η/θ varied from 0.04 to 0.12. The volumetric solids concentration C_O ranged from 0.08 to 0.25 and the *polymer powder bulk density*, ρ_{BD} from 293 to 505 kg, m^{-3}. It was checked that the legs did not run full.

It was tried to correlate v_{sett} by multiple regression analysis to all parameters possibly related to the settling legs' functioning and operation, including cycle period, polymer average particle size, polymer particle size distribution, number of settling legs in service, and properties of the reactor liquid. Eventually, the following type of correlation was withheld, only based on the reactor solids concentration and the bulk density of the polymer powder,

$$v_{sett} = \beta_O - \beta_1 C_O + \beta_2 \rho_{BD} . \tag{3.8}$$

The correlation coefficient R^2 is 0.86 (Figure 3.2).

Interesting to note is that on Figure 3.2 three data points are shown, which stem from an older plant, with settling legs of an 8 inch diameter instead of 10 inch, and which were half as long. The correlation was calculated without these data points. The position of the points confirms the validity of the formulas presented here, and of the correlation. The comparison between the two plants has also delivered useful insights in the settling of the polymer particles, that are discussed in the next Chapter 4.

This general correlation is useful by revealing the influence of the solids concentration, and by indicating the importance of the polymer bulk density.
To promote good settling, which is so important in the operation of the plant, catalysts and reactor conditions must be selected which yield the highest polymer bulk densities. The bulk density describes the settling ability of the polymer, apparently as

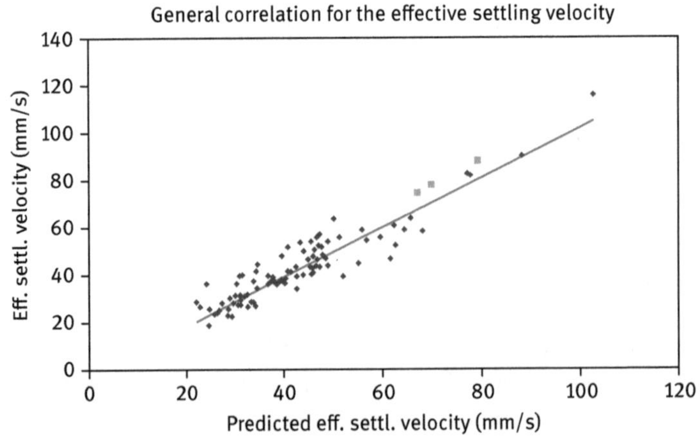

Fig. 3.2. General correlation for the Effective Settling Velocity. The three data points shown as light blue squares stem from an older plant, with settling legs of an 8 inch diameter instead of 10 inch, and half as long. The correlation was calculated without these data points.

a property which "wraps up" particle size distribution, particle shape, and particle density.
The negative value of the coefficient of the solids concentration C_0 indicates that the settling velocity is strongly reduced at higher solids concentrations, or that the settling is outspokenly *hindered*.
The general correlation should, however, not be used in practical calculations because the variance left by the correlation is too great, and the resulting errors are not acceptable.

It is important for later discussion to note that the effective settling velocity is not dependent on the cycle period T as illustrated in Figure 3.3. This confirms the assumption that v_{sett} is constant as long as the settling leg is not saturated, made in Section 3.1.1, on which the used formulas are based. The acceptance of this fact will also play a key role in the discussion of the settling in Chapter 4.
Figure 3.3 is showing the residual values of the general correlation, i.e. the differences between measured and predicted values of the effective settling velocity, plotted versus cycle period. There is no visible trend in the residuals versus cycle period, which means that the settling velocity is constant indeed, and not diminishing when the settling leg is getting more filled with time.

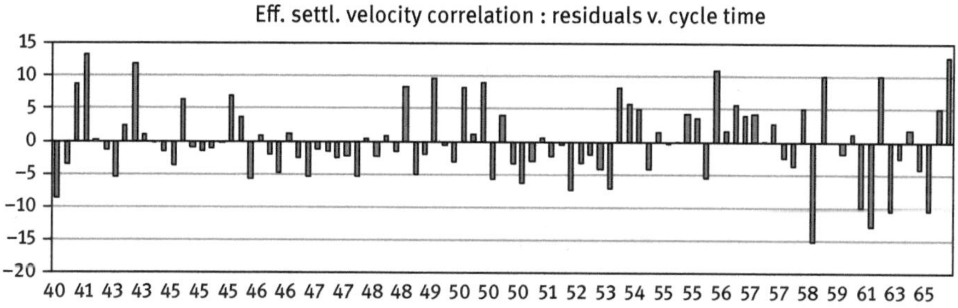

Fig. 3.3. Residuals of the general eff. settl. velocity correlation versus cycle time.

3.2.2 Specific Correlations

Correlations that are obtained separately for each catalyst family, viz. chromium oxide, Ziegler–Natta and metallocene catalysts, are more accurate, and can be used in practical models. They are of the format,

$$v_{\text{sett}} = \beta_0 - \beta_1 C_0 + \beta_2 \frac{\rho_{\text{p}} - \rho_{\text{f}}}{\rho_{\text{f}}} . \tag{3.9}$$

The coefficient for C_0 can be assumed to be valid for one catalyst family. The influence of the solids concentration due to the hindered settling is explaining the largest part

(75%) of the variance. The second parameter is a dimensionless group that is usually met with in settling problems, and explains another 5% of the variance.

In many practical cases, e.g. when only one catalyst type is involved, a correlation limited to the one parameter C_0 will be sufficient,

$$v_{sett} = \beta_0 - \beta_1 C_0 , \tag{3.10}$$

and its coefficient β_1 may be known already from other members of the catalyst family. An example of such a correlation is shown in Figure 3.4.

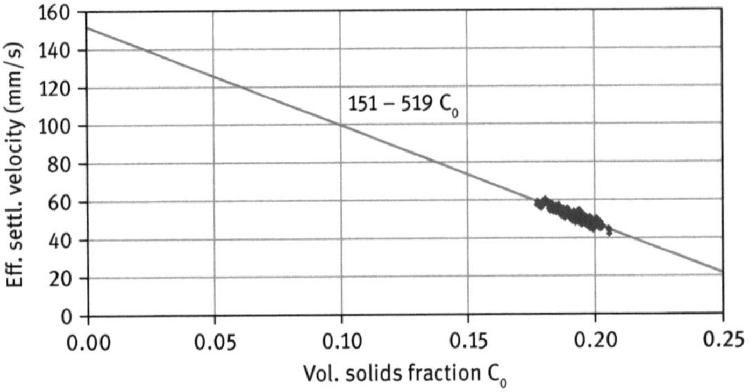

Fig. 3.4. Specific correlation for the eff. settl. velocity.

3.2.3 Correlation with the Terminal Falling Velocity

Starting from correlation (3.10), one can endeavour to generalize the specific correlations, making use of the *terminal falling velocity of the polymer particles, or Stokes' falling velocity*, v_∞, as v_{sett} should be equal to v_∞ when $C_0 \to 0$,

$$v_{sett}/v_\infty = \beta_{0,\infty} - \beta_{1,\infty} C_0 . \tag{3.11}$$

The approach does make sense, because the value of $\beta_{0,\infty}$ is indeed around 1, and it is useful in showing that the specific correlations for v_{sett} are of a type already known in the literature[5].

However, since a shape factor is wanted to make the correlation fit the specific cases of catalysts and reactor conditions, it does not bring any practical advantage over the specific correlations (3.9). The necessity of a shape factor is illustrated in Figure 3.5,

5 G. J. Kynch, *A theory of sedimentation*, Trans. Faraday Soc., 48, p. 166, 1952.

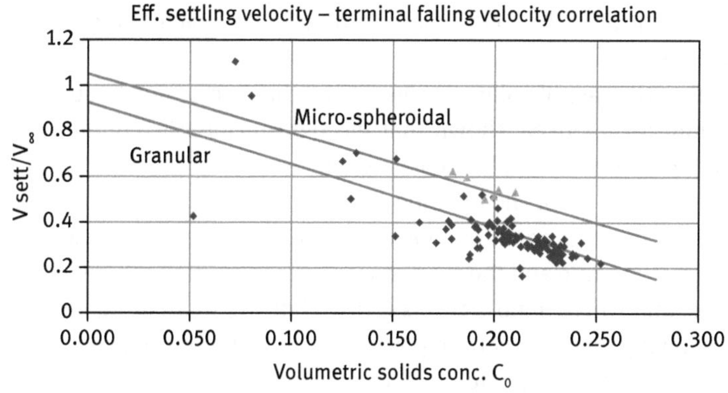

Fig. 3.5. Correlation of the eff. settl. velocity with the terminal falling velocity.

where the correlation is shown for the data set presented in Section 3.2.1, making the distinction between granular catalysts and one micro-spheroidal catalyst[6].

3.3 Control of the Solids Concentration in the Loop Reactor

After explaining the mechanisms by which the solids are removed from the loop reactor, and the quantification of the settling velocity of the polymer particles in the settling legs, the important question of how the settling leg operation is affecting the reactor operation can be answered.

3.3.1 The Solids Concentration Control Equation

The solids removal rate equation (3.6) is a mass balance for solid polymer. It describes how the produced polymer is entering the legs by displacement and sedimentation. At the same time, the volumetric output equation (3.3) should be fulfilled.
In practical operation of the reactor, it is, between certain limits, not important which volume ηL is released when the legs are opened. A higher emptied volume is automatically compensated by a lower frequency of opening. A whole range of emptied volumes can be applied without affecting the control of the loop reactor in any way. Thus it is logical to eliminate the group $\eta L/\theta$ from the two equations (3.3) and (3.6). The result is the equation,

$$P = \frac{\rho_P C_0}{1 - C_0} \left(\frac{D_R}{\rho_f} + \frac{\pi D^2}{4} v_{\text{sett}} N \right) , \tag{3.12}$$

6 A granular catalyst is supported on irregular silica particles obtained by grinding, whilst the silica particles of a micro-spheroidal catalyst are obtained by spray-drying and nearly spherical.

which is governing the control of the solids concentration in the reactor, and is thus called the *solids concentration control equation.*

It is a statement that the rate of solids removed from the reactor must equal the rate of their production, for it can be shown that the right member is equal to the rate of solids removed from the reactor. It is a mass balance valid for the steady state. Neither the settling leg length L nor the opening frequency η/θ appear in the equation. It should be stressed that this equation, however useful, is not more than a mass balance where the production rate and the reactor solids concentration are treated as independent variables, and where in no way it is verified how they can be achieved.

The equation is showing how, at a certain production rate, the settling behaviour of the polymer will set the necessary rate of diluent (isobutane, propane) to be recycled. Poor settling characteristics will impose high circulation flows and accordingly high costs in the diluent recovery section, or even limit the plant capacity. Since the settling behaviour of the polymer is mainly determined by the catalyst, as shown in Section 3.2.1, it is obvious that catalysts should not only be selected for high activity and product quality, but also for sufficient polymer settling velocities.

3.3.2 Control of the Solids Concentration

An adequate control of the solids concentration is of first importance to the stability of a loop reactor. Unfortunately, it will become clear that this control is not straightforward!

The solids concentration control equation (3.12) describes how the reactor solids concentration is controlled by the recycle diluent feed and the number of legs in service at a certain production rate. It is useful in giving the number and size of the settling legs, and the diluent flow required at that production rate.
Examples with actual values are given in Figure 3.6 and Figure 3.7, where the equation is presented as a *solids concentration control diagram* for a certain set of constant conditions, i.e. settling leg diameter, settling velocity, liquid and polymer densities, at a specified production rate. The diagram is drawn for a certain production rate, the total recycle diluent feed rate is plotted as the ordinate versus the solids concentration (here in weight fraction) in abscis. The equation (3.12) is plotted as a set of curves, one for each number N of settling legs in service.

To understand the diagram, the two mechanisms, displacement and sedimentation, should be borne in mind.

The *zero settling condition* is obtained when the settling velocity or the number of legs is put equal to zero in equation (3.12). The second term representing the sedimentation, is vanishing and only displacement is active, (curve of zero legs, Figure 3.7). In this case, the way of removing the solids and the solids control is the same as if a continuous take-off valve is used. There is a univocal relation between the recycled diluent

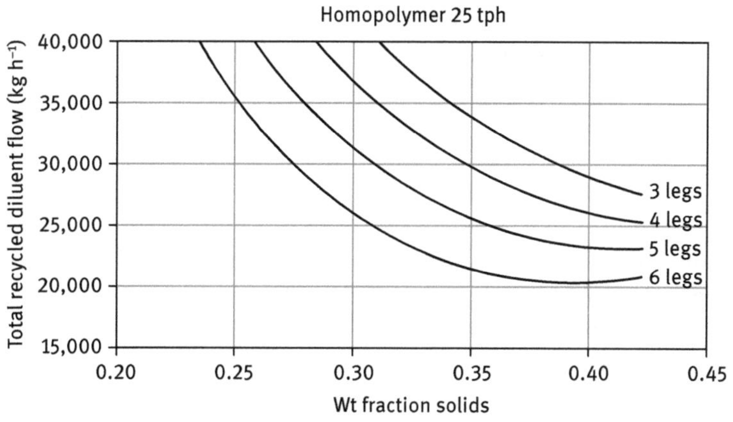

Fig. 3.6. Loop Reactor Solids Concentration Control Diagram (1).

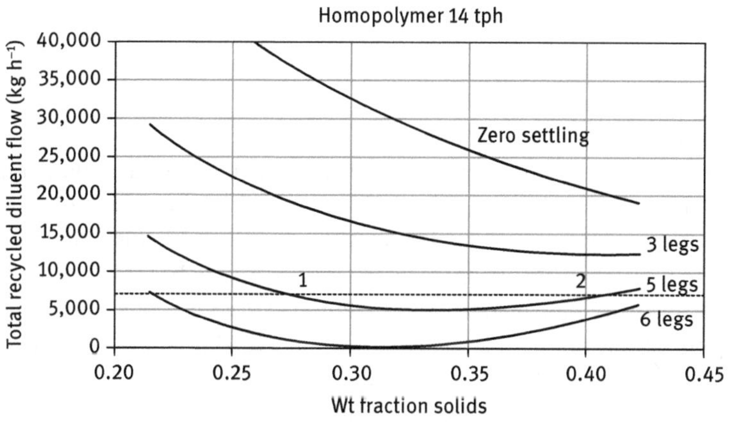

Fig. 3.7. Loop Reactor Solids Concentration Control Diagram (2).

feed and the solids concentration, the latter being "diluted by" the former. The solids concentration is simply controlled by manipulation of the recycled diluent feed rate.

The curves at higher numbers of legs are positioned lower because the same solids concentration can be maintained with lower diluent feeds, due to the additional removal of solids by the mechanism of sedimentation.

However, keeping in mind that the settling velocity is decreasing at higher solids concentrations, it is read from equation (3.6) that, at higher solids concentrations, the effect of displacement is growing and sedimentation is weakening since the slurry is thickening and the settling is more and more hindered. The curves are running closer. Eventually, all curves tend to one single point, located at the curve of zero settling, where, by extrapolation, the settling velocity has become zero, and the effect of sedimentation is lost completely. This point lies in the neighbourhood of 25 *vol%* of solids, not surprisingly where the slurry is so thick that the interaction of the particles will also hinder the pumping of the slurry, as discussed in the previous Chapter 2, Section 2.2.

3.3.3 Stability of the Solids Concentration Control

At higher recycle diluent feed rates, like for zero settling, Figure 3.7, the solids concentration is increased by decreasing the diluent feed and vice versa. This reflects the typical control, usually adopted by process computers, and fully correct where a continuous take-off valve is used to control the solids concentration, without settling legs.

However, at lower recycle diluent feeds and higher solids concentrations, the response of this concentration to the diluent feed is less and less defined: the curve of recycle diluent feed versus solids concentration is becoming less steep.
This is, as explained, due to the two mechanisms for solids removal, which are influenced in opposite directions by the solids concentration, and is more outspoken when higher areas for sedimentation (larger leg diameters or greater numbers of legs) are provided relative to the production rate.
The relation of the solids concentration to the diluent feed flow at constant production rate can even be reversed at high concentrations, where higher diluent flows are matching higher concentrations! There, it is possible to find two working points at one diluent feed flow, see points 1 and 2 at Figure 3.7. Both points do satisfy the mass balance. In point 2 a higher recycle diluent feed rate will lead to higher solids concentrations! This is not an abstraction made by extrapolation of the equations. Such areas in the solids concentration control diagram have been recorded from live plant data. An example is shown in Figure 3.8.

Speculations have been made whether these two working points can lead to instabilities, where the system would suddenly shift from one solids concentration to another. The answer is that the two working points represent just two solutions of the mass balance and do not indicate an instability of the system. The two points are located at very different solids concentrations, and thus at different catalyst residence times. To realize the same production rate at these two points would require highly different catalyst feeds and/or activities. So the system cannot just jump from the one condition to the other. The ambiguous effect of the diluent feed rate on the solids concentration shows, however, that the control of the solids concentration at relatively low diluent feeds and high solids concentrations is not straightforward. Exactly these conditions are necessary to achieve high production rates, through high catalyst residence times, and within the limits of the diluent feed capacity. There, even experienced and skilled operators may lose control from time to time, and return to safe operating areas by increasing the diluent feed rate drastically and flushing a large part of the solids out of the reactor.

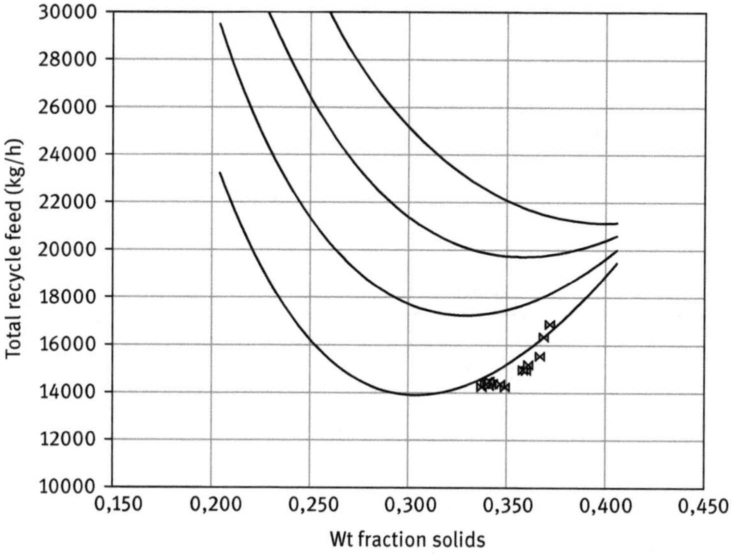

Fig. 3.8. Live plant data where the solids concentration is increased due to higher recycle diluent feed. The production rate is between 15.4 and 16.0 *tph*. There are 6 settling legs in service. Live data points are plotted on the according Solids Concentration Control Diagram. During a 3 hours' interval the recycle diluent feed rate is increased from 14.4 to 16.9 *tph*. The solids concentration is increasing from 34 to 37 *wt*%.

The question of the stability of the solids control is thus raised inevitably.
To study this stability, one has to take into account how the production rate is achieved, which up to here has been avoided, and kept for later chapters.
The production rate P must indeed be explicited in the solids concentration control equation (3.12). It is, among others, dependent on the solids concentration itself. A simple example may be presented here, where the catalyst activity $\overline{A}$ is supposed to be constant, and thus a simple expression is obtained for the production rate[7],

$$P = (F_{\text{cat}}\overline{A}V_{\text{R}}C_{\text{O}}\rho_{\text{p}})^{0.5} = \frac{\rho_{\text{p}}C_{\text{O}}}{1 - C_{\text{O}}}\left[\frac{D_{\text{R}}}{\rho_{\text{f}}} + \frac{\pi D^2}{4}(\beta_{\text{O}} - \beta_1 C_{\text{O}})N\right]. \tag{3.13}$$

This case is plotted, making use of realistic values, in Figure 3.9, where equation (3.13) is presented as two curves, one for the rate of solids produced, the left member of the equation, and one for the rate of solids removed by the settling legs, the right member, both plotted versus the solids concentration. Working points are found at the intersections of the curves.
Point 1 is intrinsically stable. If by a stochastic deviation the solids concentration should be increased, the resulting removal rate of solids would be higher than the

7 See below equation, (7.13).

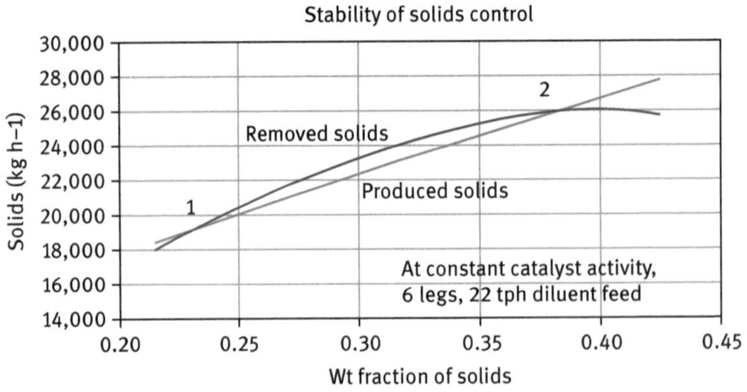

Fig. 3.9. Stability of the solids concentration control in the loop reactor.

production rate, and thus the system will return to the working point. Working point 2 is much closer to real life operating conditions when a loop reactor is well loaded to obtain a high and economical production rate. Unfortunately, this working point 2 is not intrinsically stable! A deviation toward higher solids concentrations will there effect a production rate of solids that is higher than the removal rate of solids, and thus strengthen the deviation toward still higher concentrations.

The instability problem is not limited to this hypothetical case of constant catalyst activity. More precise expressions for the catalyst activity do not allow to lift the instability. Catalyst activity will be shown to be rather decreasing with catalyst residence time and thus with solids concentration, but not enough to reverse the balance between solids production and removal. There is no other solution to this problem than to welcome and appreciate, also here, the skills and enthusiasm of a well motivated reactor operator.

Notations

$\overline{\mathcal{A}}$	catalyst activity: the quantity of polymer produced by a quantity of catalyst present in the reactor per unit time, (*kg* polymer/*kg* catalyst, *s*).
C_0	volumetric concentration of solids in the reactor, (–).
C_T	average volumetric solids concentration in the total emptied volume fractions of all settling legs in service at the end of each cycle period *T*, (–).
D	settling leg internal diameter, (*m*).
D_R	total flow of recycled diluent (reactor liquid, or flash gas) leaving the reactor, (kg, s^{-1}).
F_{cat}	catalyst feed to the reactor, (kg, s^{-1}).

N	number of settling legs in service, (–).
L	total length of the settling leg, (m).
P	polymer production rate: the total quantity of polymer produced in the reactor per unit of time, (kg polymer, s^{-1}).
T	cycle period: period between two consecutive openings of the same settling leg, (s).
V_R	reactor volume, (m^3).
v_{sett}	effective settling velocity: average downward velocity by which settling polymer particles enter the emptied fraction of the settling leg, (m, s^{-1}).
v_∞	terminal falling velocity of the polymer particles in the reactor liquid, (m, s^{-1}).
w	weight fraction of solids in the slurry discharged by the settling legs, (kg, kg^{-1}).
η	emptied fraction: fraction of the settling leg's volume, emptied at every opening of the leg, (–).
η/θ	opening frequency: number of times per second that the total leg volume is discharged, (s^{-1}).
θ	opening period: period between two consecutive openings of any settling leg, (s).
ρ_{BD}	bulk density of the polymer powder produced in the reactor, (kg, m^{-3}).
ρ_f	density of the liquid in the reactor, (kg, m^{-3}).
ρ_p	density of the polymer in the reactor, (kg, m^{-3}).
ΣF_i	sum of all feeds to the reactor, (kg, s^{-1}).

4 The Settling of the Polymer in the Settling Legs

The model for the functioning of the settling legs as described in the previous Chapter 3 is merely empirical, since v_{sett} has been defined and calculated purely empirically, without any reference to what is actually happening in the legs. Still, it is sufficient to allow the operation of the settling legs. The reader may leave unread what follows, and direct himself to the next chapter, where the subject of the chemical reaction is started, without losing the thread of the discussion.

However, this chapter offers a more profound understanding of the settling of the polymer particles in the legs, and will justify many of the assumptions made in the previous chapter. It will make the reader feel more comfortable when applying the model, more skilled when designing a settling leg system, and even pleased with the aesthetics of a nice mathematical deduction.

The need for this better understanding occurred to the author when comparing data from two different plants, a few thousand kilometers apart, and of quite different dimensions. This data was taken from production runs where the two plants made exactly the same product, from the same catalyst and at identical reactor conditions. The settling legs in the one plant owned a 10 inch diameter. In the other plant, the settling legs showed an upper part connected to the loop reactor vessel of a smaller 8 inch diameter, whilst the lower part had a 10 inch diameter, just like in the first plant. The upper part of the legs, including valves and connections to the reactor vessel, made up more than half of their total length. Intuitively, it was expected that the more narrow upper part in the second plant would there restrict the settling rate of the polymer. Surprisingly, the performance of the legs was identical in both plants[1].

An explanation of the settling process in the legs is offered by the sedimentation theory of G. J. Kynch, published in 1952. Kynch's theory has been criticized in the past[2], but is fully applicable to incompressible particles, that do not absorb any liquid, as has been assumed for the polymer particles here, Section 3.1.1.

Accepting that some polymer wil be collected at the bottom of the leg, with a uniform maximal solids concentration (related to the polymer's bulk density), it will first be shown from Kynch's theory that this bottom layer will grow at a constant speed.

---

1 This finding was described in US Patent Application US2005/0038206 A1, Feb. 17, 2005, by Ingrid Melaaen, Paul Allemeersch et al.

2 G. J. Kynch, *A theory of sedimentation*, Trans. Faraday Soc., 48, p. 166, 1952. Kynch's article surprised the sedimentation technologists, who, since 1916, had relied in their design on an obscure article from Coe and Clevenger of the US Bureau of Mines. Kynch was a mathematician at the University of Birmingham, UK. His use of differential equations must have been new to the chemical engineers from the 1950s. His contribution was welcomed at first, but later criticized. Not rightly so, because where his theory was said to fail, particles are not well defined. Colloidal particles that are met in the thickening of sewage sludges and flocculated matters change in nature throughout the process, by compression and loss of liquid.

Then, keeping in mind the experimental fact that the flux of polymer entering the leg is constant in time, the whole concentration profile in the leg, on top of the bottom layer, will be found to be uniform and equal to the reactor concentration. This conclusion will be confirmed by the form of the correlation for the effective settling velocity. A concise discussion of settling legs with two diameters will also be given.

4.1 Description and Definitions

The same simplifying assumptions are made as in the previous chapter, Section 3.1.1.

1. Steady state conditions.
2. Production rate and reactor solids concentration are treated as independent variables.
3. The reactor slurry is considered as consisting of two phases, pure polymer and pure liquid, in which polymer density and reactor liquid density are constants.

The settling leg is considered at rest, with only sedimentation going on. The displacement that occurred at the last opening of the leg has been immediate, taking a zero time interval, and no turbulence consequential to the opening is taken into account.

Five different zones, or interfaces, are assumed to exist in the emptied fraction η of the leg's volume, which in the previous chapter has not been described in detail (Figure 4.1).

- Zone A: the solids concentration is equal to the reactor concentration C_0 by backmixing from the reactor.
- Interface B: the cross-section where the backmixing has stopped, and from where on particles make a downward movement by sedimentation.
- Zone C: a zone of continuous settling, showing a continuous concentration profile $C(x, t)$.

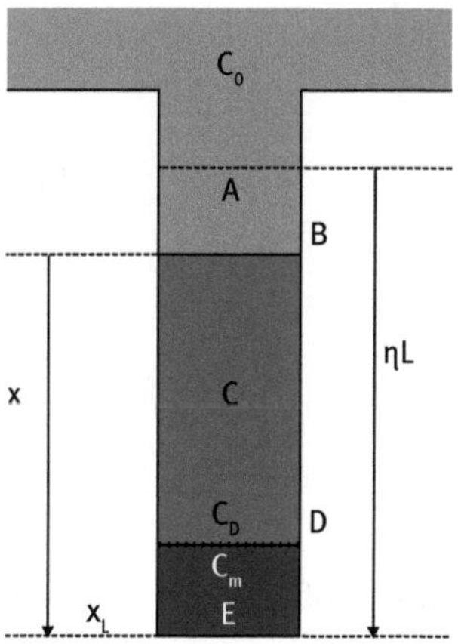

Fig. 4.1. The settling of the polymer in the settling legs: definitions.

- Interface D: the interface between zones C and E.
- Zone E: a zone filled with a thickened slurry at the maximal solids concentration C_m.

The concentration profile $C(x, t)$ in the zone C of the leg is a function of the time t during the cycle period, and is supposed to be unidimensional. x is the position on the single coordinate axis indicating a position along the leg, $x = 0$ at the interface B, and $x = x_L$ at the bottom of the leg. At the opening of the leg, the whole emptied volume of the leg is at the uniform reactor solids concentration, or $C(x, t = 0) = C_0$.

4.2 Kynch's Sedimentation Theory

This section is a mere transcription of Kynch's theory, so all tribute should be paid to him. Two basic elements from the theory are reproduced and applied to the zones C and E.

Let S be the flux of polymer particles passing a cross section of the settling leg by sedimentation (m^3 polymer, m^{-2}, s^{-1}).
Let v be the settling velocity of the polymer particles (m, s^{-1}).

The important assumption is made that the settling velocity v is only a function of the solids concentration C, $v = v(C)$, which means that the settling velocity is determined by the immediate environment of the particle, and not influenced e.g. by boundary conditions from the walls of the settling leg tube.
Then the polymer flux S, which is equal to $C \cdot v$, is also only a function of C,

$$S = C \cdot v(C) \,.$$

A mass balance is written for a thin slab of the zones C, D and E with thickness $\mathrm{d}x$,

$$\frac{\partial C}{\partial t}\mathrm{d}t\frac{\pi D^2}{4}\mathrm{d}x = S(x)\frac{\pi D^2}{4}\mathrm{d}t - S(x+\mathrm{d}x)\frac{\pi D^2}{4}\mathrm{d}t \,,$$

where D is the internal diameter of the settling leg.
Hence,

$$\frac{\partial C}{\partial t} = \frac{S(x) - S(x+\mathrm{d}x)}{\mathrm{d}x} = -\frac{\partial S}{\partial x} \,, \tag{4.1}$$

or, since S is only a function of C,

$$\frac{\partial C}{\partial t} + \frac{\mathrm{d}S}{\mathrm{d}C}\frac{\partial C}{\partial x} = 0 \,. \tag{4.2}$$

Finally, since $S = C \cdot v$,

$$\frac{\partial C}{\partial t} + \left(v + C\frac{\mathrm{d}v}{\mathrm{d}C}\right)\frac{\partial C}{\partial x} = 0 \,. \tag{4.3}$$

Equation (4.3) allows, in principle, to solve the concentration profile $C(x, t)$ if the settling velocity $v(C)$ is known.

During the process, more and more particles will settle in the leg, so $C(x, t)$ will, in general, increase with time.

A cross-section of the slurry at a position x can be referred to as a *layer* of the slurry at x. But a layer, independent of x, can also be imagined, where the solids concentration is equal to a certain value of C. It is interesting to follow such a layer of a constant concentration C.

A layer of constant concentration C is defined by $\mathrm{d}C = 0$, or,

$$\frac{\partial C}{\partial t}\mathrm{d}t + \frac{\partial C}{\partial x}\mathrm{d}x = 0\,,$$

where x and t are the coordinates in place and time of the layer of concentration C. Hence,

$$\frac{\mathrm{d}x}{\mathrm{d}t} = \frac{-\frac{\partial C}{\partial t}}{\frac{\partial C}{\partial x}}\,,$$

and, using equation (4.1),

$$\frac{\mathrm{d}x}{\mathrm{d}t} = \frac{\mathrm{d}S}{\mathrm{d}C}\,. \tag{4.4}$$

The latter equation (4.4) leads to the important conclusion that, since S is only a function of C, each layer of a constant concentration C will propagate through the leg with a constant velocity $\mathrm{d}S/\mathrm{d}C$.

E.g. in a concentrated slurry where S is decreasing with higher concentrations, $\mathrm{d}S/\mathrm{d}C < 0$, a layer of such concentration will move upwards through the leg, because the flux S leaving it is lower compared to the flux S entering it from the superjacent layer.

It is important to note that equation (4.4) is very general and is founded on only one assumption, i.e. that the settling velocity v is only a function of the concentration C.

Kynch's theory allows to construct the polymer flux curve $S = C\cdot v(C)$ from a single sedimentation experiment where the position of the upper layer of the solids is noted as a function of time[3], and thus to obtain $v(C) = S/C$ and $\mathrm{d}x/\mathrm{d}t = \mathrm{d}S/\mathrm{d}C$.

This knowledge allows to characterize the sedimentation behaviour of the slurry[4].

Some slurries exhibit a discontinuity in their concentration profiles, which is explained by Kynch's theory. The upward propagation speed of a layer in a concentrated

3 How to do this, is described by Kynch, ibid.

4 Lab-scale sedimentation tests for polymer from newly developed catalyst types seem recommended to verify its settling behaviour. However, when looking into the laws of downscaling, based on the dimensionless equations that are governing the settling, meaningful lab-scale sedimentation tests are very difficult to design. The polymer particles produced on lab-scale should be identical to those from full-scale operation, which is never the case. The density of the liquid, and possibly its viscosity (in the transient range between turbulent and laminar flow), should be the same as for the full-scale diluent, viz. a light hydrocarbon at about 100 °C. The author has no knowledge of any successful lab-scale sedimentation tests on such polymers.

slurry where $\mathrm{d}S/\mathrm{d}C < 0$, will be higher than the speed of the superjacent layers when $\mathrm{d}S/\mathrm{d}C$ in the former layer is more negative than in the latter. Those will instantly be "caught up" by the faster underlying layer. It means that all concentrations where $\mathrm{d}S/\mathrm{d}C < 0$ and $\mathrm{d}^2S/\mathrm{d}C^2 < 0$, (or $\mathrm{d}S/\mathrm{d}C$ larger in absolute value, but more negative), will be non-existent, and a discontinuity in the concentration profile will occur. An example of a such flux curve is shown in Figure 4.3, below, at the end of Section 4.4, where $S(C)$ is a parabolic curve. When the sedimentation starts at a concentration value C_0 at the right side of the curve, where $\mathrm{d}S/\mathrm{d}C < 0$, the slope will be more negative at all concentrations $C_0 < C < C_\mathrm{m}$ between C_0 and the maximum concentration C_m of the packed solids. Those intermediate concentrations will be non-existent and a discontinuity will occur from C_0 to C_m.

4.3 Growth Rate of the Zone of Thickened Slurry

It is now assumed that a zone of thickened slurry with the maximal concentration C_m is existing at the bottom of the leg, viz. that the zone E is existing.
What follows is proving from Kynch's theory that the growth rate of that zone, or u_D, *the upward velocity of the interface D*, is a constant. By definition, u_D is,

$$u_D = -\frac{\mathrm{d}x_D}{\mathrm{d}t} ,$$

with $x_D(t = 0) = x_L$, since the thickness of the bottom layer is zero at the start of the cycle period.

It is supposed at first that the concentration profile $C(x, t)$ is continuous at x_D. In that case the concentration at x_D is C_m, and the general case, equation (4.4), is valid,

$$u_D = -\frac{\mathrm{d}x_D}{\mathrm{d}t} = -\left(\frac{\mathrm{d}S}{\mathrm{d}C}\right)_{C=C_\mathrm{m}} . \tag{4.5}$$

Hence, u_D is a constant, because C_m is considered to be a constant for a certain type of polymer powder.

Next it is considered that a discontinuity is existing at x_D, because many slurries show such settling behaviour.
In this case, $C = C_\mathrm{m}$ at $x = x_D^+$, and $C = C_D(x_D, t)$ at $x = x_D^-$.
The reasoning that is now followed has been indicated, though not so well explained, by Fitch[5].
The layer at $x = x_D^-$ with concentration C_D is considered, and two statements are made about u_D, i.e. the speed at which this layer, together with the layer at $x = x_D^+$, is propagating.

5 Bryant Fitch, *Current theory and thickener design*, Industrial and Engineering Chemistry, 58, p. 18, Oct. 1966.

First, a mass balance for the growth of the thickened zone E is written, stating that the build-up of the zone is effected by the flux of polymer settling at x_D,

$$-\frac{\mathrm{d}x_D}{\mathrm{d}t}\mathrm{d}t\frac{\pi D^2}{4}(C_\mathrm{m} - C_D) = C_D v(C_D)\frac{\pi D^2}{4}\mathrm{d}t\,.$$

Hence,

$$u_D = -\frac{\mathrm{d}x_D}{\mathrm{d}t} = \frac{C_D v(C_D)}{C_\mathrm{m} - C_D} = \frac{S(C_D)}{C_\mathrm{m} - C_D}\,. \tag{4.6}$$

Secondly, the general condition, equation (4.4), should be fulfilled, and the layer at $x = x_D^-$ with concentration C_D is propagating with the speed,

$$u_D = -\frac{\mathrm{d}x_D}{\mathrm{d}t} = -\left(\frac{\mathrm{d}S}{\mathrm{d}C}\right)_{C=C_D}\,.$$

The two expressions for u_D should be equal and thus, the variable C_D should fulfil,

$$\left(\frac{\mathrm{d}S}{\mathrm{d}C}\right)_{C=C_D} = -\frac{S(C_D)}{C_\mathrm{m} - C_D}\,. \tag{4.7}$$

The latter expression (4.7) must be valid for all slurries where a discontinuity occurs at the interface D, no matter what settling behaviour $S(C)$ is shown by the slurry.
The concentration C_D has been seen as a variable so far, but, if a variable, then the flux S(C) must be a solution of,

$$\frac{\mathrm{d}S}{S} = -\frac{\mathrm{d}C}{(C_\mathrm{m} - C)}\,,$$

with $S(0) = 0$.
It would follow that S(C) can only be one definite function for a certain value of C_m, which does not hold for all slurries in general, hence C_D must be a constant, and so must u_D.

4.4 Settling Behaviour of the Polymer

It will now be shown that, throughout the cycle period, the slurry in the leg is composed only of a zone C of a constant solids concentration C_0, which is the reactor concentration, and a zone E of the maximal concentration C_m, with a discontinuity at the interface D. The way to do this is based on the experimental fact, mentioned in Section 3.2.1 of the previous Chapter 3, and illustrated in Figure 3.3, that the effective settling velocity v_sett is independent of the cycle period T, and thus that the polymer flux entering the settling leg, viz. the emptied volume of the settling leg, is constant in time.

The settling behaviour of the polymer in the settling legs should be inferred from the following data.

- The flux of polymer entering the leg at interface B is constant in time (at fixed reactor conditions), ref. Section 3.2.1, and equal to,

$$S(x=0,t)=S(C_0)=v(C_0)C_0=c\,.$$

- By its definition in Section 3.1.4, v_{sett} can be identified with $v(C_0)$, $v_{\text{sett}}\equiv v(C_0)$, and from experiment, Section 3.2.2, equation (3.10), $v(C_0)$ is thus known as,

$$v(C_0)\equiv v_{\text{sett}}=\beta_0-\beta_1 C_0\,.$$

- The concentration profile of the solids $C(x,t)$ near the settled zone E is either continuous or discontinuous. From the previous Section 4.3, equation (4.5), it is known that in the former case the propagation speed u_D of the interface D is,

$$u_D=-\left(\frac{\mathrm{d}S}{\mathrm{d}C}\right)_{C=C_D=C_\mathrm{m}}=c\,,$$

  whilst in the latter case, equation (4.6),

$$u_D=\frac{S(C_D)}{C_\mathrm{m}-C_D}=c\,.$$

  In both cases, u_D and C_D are constant in time.
- The boundary condition is valid that at the start of the settling period T, all concentrations are equal to the reactor concentration,

$$C(x,t=0)=C_0\,.$$

The emptied volume of the settling leg is composed of the zones C and E. The flux of polymer entering these zones at interface B is constant in time, so the rate at which the total polymer content of the zones C and E is increasing, must be equal to the polymer flux at B and constant in time,

$$\frac{\pi D^2}{4}S(C_0)=\frac{\mathrm{d}}{\mathrm{d}t}\left(\int_{x=0}^{x=x_D}\frac{\pi D^2}{4}C(x,t)\mathrm{d}x\right)+\frac{\mathrm{d}}{\mathrm{d}t}\left(\int_{x=x_D}^{x=x_L}\frac{\pi D^2}{4}C_\mathrm{m}\mathrm{d}x\right).$$

Or,

$$S(C_0)=\frac{\mathrm{d}}{\mathrm{d}t}\left(\int_{x=0}^{x=x_D}C(x,t)\mathrm{d}x\right)+\frac{\mathrm{d}}{\mathrm{d}t}[C_\mathrm{m}(x_L-x_D)]\,,$$

and

$$S(C_0)=\frac{\mathrm{d}}{\mathrm{d}t}\left(\int_{x=0}^{x=x_D}C(x,t)\mathrm{d}x\right)+u_D C_\mathrm{m}\,. \tag{4.8}$$

The right term of the right member of the equation is a constant.

Hence the left term must be a constant as well,

$$\frac{\mathrm{d}}{\mathrm{d}t}\left(\int\limits_{x=0}^{x=x_D} C(x,t)\mathrm{d}x\right) = c\,.$$

The time variable t can be replaced by the variable x_D, since $x_D = x_L - u_D t$ and t and $\mathrm{d}t$ can be substituted by $t = (x_L - x_D)/u_D$ and $\mathrm{d}t = -\mathrm{d}x_D/u_D$ respectively,

$$-u_D\frac{\mathrm{d}}{\mathrm{d}x_D}\left(\int\limits_{x=0}^{x=x_D} C\left(x, \frac{x_L - x_D}{u_D}\right)\mathrm{d}x\right) = c\,.$$

Or, working out the derivative[6],

$$-u_D\int\limits_{x=0}^{x=x_D}\frac{\partial C\left(x, \frac{x_L-x_D}{u_D}\right)}{\partial x_D}\mathrm{d}x - u_D C\left(x = x_D, \frac{x_L - x_D}{u_D}\right) = c\,,$$

and,

$$-u_D\int\limits_{x=0}^{x=x_D}\frac{\partial C\left(x, \frac{x_L-x_D}{u_D}\right)}{\partial x_D}\mathrm{d}x - u_D C_D = c\,.$$

The right term of the latter expression is a constant, so the left term must be a constant too, for all values of x_D or t. Turning back to the variable t, this left term is equal to

$$-u_D\int\limits_{x=0}^{x=x_D}\frac{\partial C\left(x, \frac{x_L-x_D}{u_D}\right)}{\partial x_D}\mathrm{d}x = \int\limits_{x=0}^{x=x_D}\frac{\partial C(x,t)}{\partial t}\mathrm{d}x = c\,,$$

and applying equation (4.1) from Kynch's theory,

$$\int\limits_{x=0}^{x=x_D}\frac{\partial C(x,t)}{\partial t}\mathrm{d}x = -\int\limits_{x=0}^{x=x_D}\frac{\partial S(x,t)}{\partial x}\mathrm{d}x = c\,.$$

The value of this constant term can be obtained at any value of t, and thus also from the conditions at $t = 0$. There the concentration $C(x, t = 0) = C_0$ for all x, and thus $\partial S(x,t)/\partial x$ is zero for all x. It follows that this term is equal to zero.

6 An integral with a variable limit is differentiated to that limit as,

$$\frac{\mathrm{d}}{\mathrm{d}t}\left(\int\limits_a^t f(x)\mathrm{d}x\right) = f(x = t)\,.$$

When the integrated function is also dependent on that limit, the derivative is,

$$\frac{\mathrm{d}}{\mathrm{d}t}\left(\int\limits_a^t f(x,t)\mathrm{d}x\right) = \int\limits_a^t \frac{\partial f}{\partial t}\mathrm{d}x + f(x = t, t)\,.$$

The variable limit is here x_D.

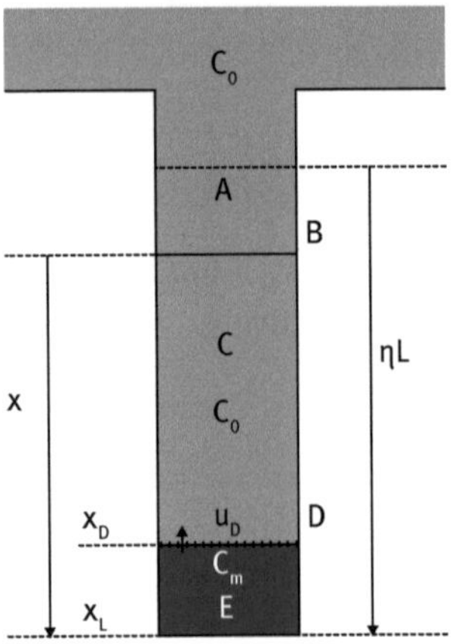

Fig. 4.2. The concentration profile of the polymer in the settling legs as derived from Kynch's theory and the experimental data.

The information gained can now be filled in in equation (4.8),

$$S(C_0) = 0 - u_D C_D + u_D C_m ,$$

or,

$$S(C_0) = u_D (C_m - C_D) . \tag{4.9}$$

This simple result has been obtained without knowing yet whether the concentration profile near interface D is continuous or discontinuous.
If continuous, $C_D = C_m$ and $S(C_0) = 0$, which is not true.
So there must be a discontinuity at the interface D. Then C_D is constant in time, and necessarily equal to $C_D(x_D, t) = C_D(x_D, t = 0) = C_0$, and equation (4.9) becomes,

$$u_D = \frac{S(C_0)}{(C_m - C_0)} .$$

This is in complete agreement with equation (4.6) for the case of a discontinuity at D with $C_D = C_0$!

The settling behaviour of the slurry has thus been detected. At the interface D, the concentration jumps from C_m to C_0. To allow that discontinuity it must be that for all intermediate concentrations C between C_0 and C_m,

$$\left(\frac{\partial S}{\partial C}\right)_{C_0<C<C_m} < \left(\frac{\partial S}{\partial C}\right)_{C=C_0} < 0 ,$$

so that layers of these intermediate concentrations are inexistent. Throughout the cycle period, the slurry in the leg must be composed only of a zone C of a constant solids concentration C_0, which is the reactor concentration, and a zone E of the maximal concentration C_m, with a discontinuity at the interface D, as presented in Figure 4.2.

And indeed, this deduction of the settling behaviour is confirmed by the experimental correlation for the settling velocity v.

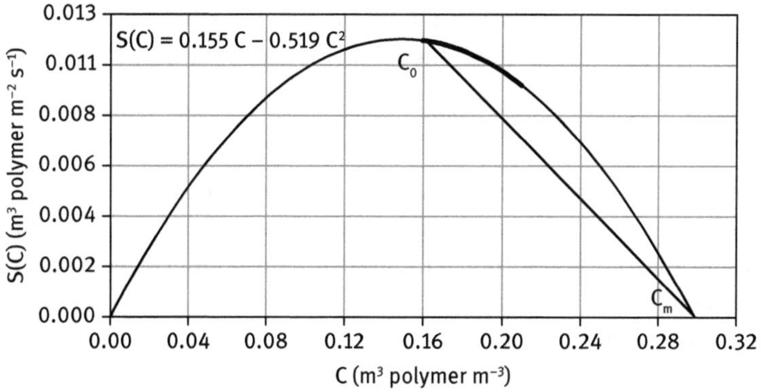

Fig. 4.3. The polymer flux curve $S(C)$ as derived from the empirical settling velocity $v_{\text{sett}}(C)$. The bold part of the flux curve stems from experimental data.

This correlation $v(C_0)$, equation (3.10), has been obtained from different stable reactor conditions at different solids concentrations C_0. It is thus part of a general expression for $v(C)$, and so reveals part of the polymer flux curve,

$$S(C) = v(C)C = (\beta_0 - \beta_1 C)C = \beta_0 C - \beta_1 C^2 .$$

The polymer flux curve $S(C)$ is thus a parabolic function of C. In the region of interest the experimental values of the coefficients are such that $C > \beta_0/2\beta_1$, and will, according to Kynch's theory, effect the above described settling behaviour and discontinuity. Fig 4.3 is showing a realistic example of the flux curve and the resulting discontinuity in the concentration profile.

The detection of the settling behaviour of the polymer is a nice example of how valuable information can be obtained directly from full scale plant data.

4.5 Maximum Filled Fraction of the Settling Legs

The concentration profile that has now been established, explains why the assumption and necessary condition, made in Section 3.1 above, that the solids concentration in the upper volume fraction $(1 - \eta)$ of the leg should always be equal to the solids concentration in the reactor C_0, is so easily fulfilled in practical reactor operation at widely varying values of η.

Still, for proper operation of the reactor, a minimum value of η should be respected, lest the legs run full with packed polymer. Defining the *filled fraction* η_f as the fraction of the leg volume that is, at the end of the cycle period T, filled with polymer packed at the maximal concentration C_m, the emptied fraction η must be larger than η_f.

The maximum volumetric solids concentration, C_m, can be estimated as equal to the volumetric solids concentration of the settled fluff, on which the bulk density ρ_{BD} is determined in the lab,

$$C_m = \frac{\rho_{BD}}{\rho}\,, \tag{4.10}$$

where ρ is the density of the polymer, as measured in the lab at room temperature. Sedimentation will thicken the slurry in the leg from the reactor slurry concentration to the average concentration C_T, see Section 3.1.2, and form the bottom layer of packed polymer in the leg. The volume of polymer deposited during the cycle period T can now be expressed in terms of η_f since the concentration profile in the leg is known,

$$\frac{\pi D^2}{4} C_0 v_{sett} T = \eta \frac{\pi D^2}{4} L(C_T - C_0) = \eta_f \frac{\pi D^2}{4} L(C_m - C_0)\,.$$

Whence the filled fraction is derived,

$$\eta_f = \eta \frac{C_T - C_0}{C_m - C_0}\,, \tag{4.11}$$

as well as the simple condition that the legs should not run full,

$$\frac{\eta_f}{\eta} = \frac{C_T - C_0}{C_m - C_0} \leq 1\,. \tag{4.12}$$

4.6 Settling Legs with a Change in Diameter

Settling legs may have a smaller diameter in the upper, and a larger diameter in the lower part. Such design allows to conserve the existing reactor nozzles and top valves in case of a reactor revamp, or to save costs by using smaller size nozzles and valves in case of a new design. The sedimentation capacity of the legs is, within limits, the same, as if the whole leg were constructed with the largest diameter.

As mentioned in the introduction to this chapter, data obtained from two reactors under identical conditions, the one with a uniform leg diameter of 10 inch, and the other with a diameter change from 8 inch to 10 inch, has shown that the settling leg sedimentation capacity is basically the same, and is determined by the lower and larger 10 inch diameter. The same values of $v_{sett} \equiv v(C_0)$, based on the largest diameter D_2, were recorded. This fact can be explained in view of the results obtained in this chapter.

Considering the same zones as in Section 4.1 above, zone C is now divided into two zones $C1$ and $C2$ with different diameters D_1 and D_2, Figure 4.4.
In zones $C2$-D-E the same situation is found as described in this chapter. The rate of settling polymer is determined by the diameter D_2 and the polymer flux $S(C_0)$.

At the opening of the leg, settling will start at the bottom of the leg with a flux $S(C_0)$. This settling rate is the same as if the leg had owned a uniform diameter D_2, and will, between limits, be maintained throughout the whole leg. At the diameter change x_1, this rate will extract solids from zone $C1$, which at $t = 0$ is also at concentration C_0. This extraction of solids at a rate $S(C_0)(\pi D_2^2/4)$ will reduce the solids

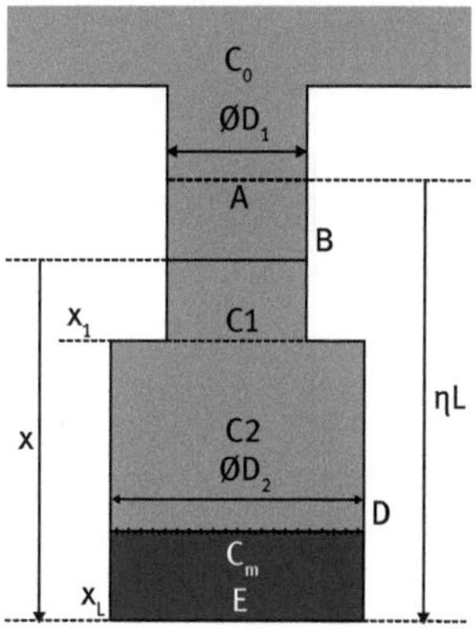

Fig. 4.4. Settling leg with a change in diameter.

concentration in the zone $C1$ toward a concentration C_1, but at the same time enhance the flux in zone $C1$ because $S(C_1) > S(C_0)$ when $C_1 < C_0$. The lower concentration C_1 will reach a stable value when continuity in mass flow is established, and $S(C_1)(\pi D_1^2/4) = S(C_0)(\pi D_2^2/4)$.

Or, due to the sudden diameter change at $x = x_1$, a discontinuity is expected to occur in the solids concentrations, where $C = C_0$ at $x = x_1^+$, and $C = C_1$ at $x = x_1^-$. The polymer flux must show a discontinuity to conserve the continuity in mass flow, and the concentration C_1 in the more narrow upper part of the leg, must fulfil the condition that,

$$S(C_1) = \left(\frac{D_2}{D_1}\right)^2 S(C_0)\,. \tag{4.13}$$

Together with the empirical expression for the polymer flux, $S(C) = \beta_0 C - \beta_1 C^2$, a quadratic equation is thus obtained wherefrom the solids concentration C_1 in zone $C1$ can be calculated.

This description explains why the sedimentation capacity of a leg with a change in diameter has been found to be the same as if this leg had owned the largest diameter over its full length. However, there must be restrictions to this phenomenon. There will be no solution for C_1 if the flux curve does not allow a high enough flux. And the zone $C1$ can only be at a uniform concentration C_1, if the flux curve allows that discontinuity, which means that for all intermediate concentrations C between C_1 and C_0, should be valid that,

$$\left(\frac{\partial S}{\partial C}\right)_{C_1<C<C_0} < \left(\frac{\partial S}{\partial C}\right)_{C=C_1} < 0\,.$$

If these two conditions are not met, the smaller diameter zone will be restrictive to the polymer flow[7].

7 This approach may be conservative, because other mechanisms than purely gravitational settling, like convection and collisions, may enhance the flux in the upper part of the leg.

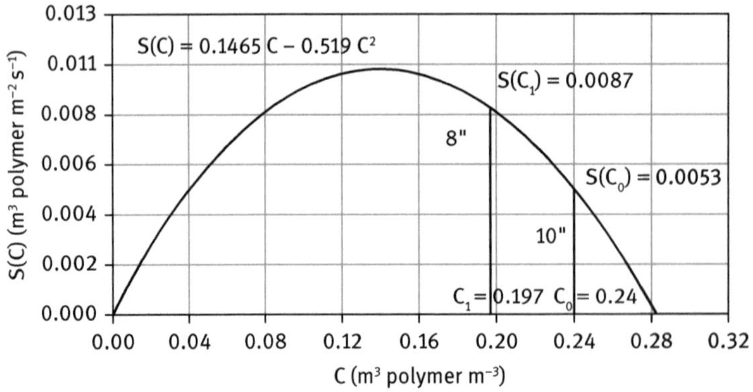

Fig. 4.5. Polymer flux curve in a realistic case of settling legs with a change in diameter.

Knowing the empirical polymer flux curve to be a parabola, the above conditions can be stated in simple words by saying that C_1 should be located at a higher point of the flux curve than C_0, and to the right of the vertex of the parabola, where the slope is negative. Figure 4.5 shows a realistic example of the conditions in a leg with a change in diameter, where this is the case indeed. The polymer flux in the zone of smaller diameter $C1$ does match the condition imposed by equation (4.13). The shape of the flux curve between $S(C_1)$ and $S(C_0)$ makes that the concentration in zone $C1$ will be uniform at C_1.

In case of legs of two diameters, the condition to prevent the legs from running full, equation (4.12), becomes more stringent, because here the lower part of the legs, at the larger diameter, should not run full, lest the legs will lose a large part of their capacity.
If η_2 is defined as the fraction of the leg volume with the larger diameter D_2,

$$\eta_2 = \frac{(\pi D_2^2/4)L_2}{(\pi D_1^2/4)L_1 + (\pi D_2^2/4)L_2},$$

then the condition will be that,

$$\frac{\eta_f}{\eta} = \frac{C_T - C_0}{C_m - C_0} \leq \frac{\eta_2}{\eta}. \tag{4.14}$$

The equations discussed in the previous Chapter 3 can be adapted for settling legs with a change in diameter, if, besides the case of normal operation, are also considered the cases where the narrow upper part of the leg becomes restrictive to the flow, and where the lower part of the leg runs full. Which case will prevail depends on the solids concentration and on the settling behaviour of the polymer. This means that the curve of solids removal, mentioned in Section 3.3.3, will show an inflection where a different case becomes valid, and there, the stability of the solids concentration may be doubtful. Therefore, settling legs with a change in diameter are not recommended.

Notations

C_0	volumetric concentration of solids in the reactor, (–).
C_m	maximal volumetric solids concentration of the solids possible in the slurry, and reached by the settled polymer particles in the bottom of the leg, (–).
C_T	average volumetric solids concentration in the total of all emptied volume fractions of all settling legs in service at the end of each cycle period T, (–).
$C(x, t)$	volumetric solids concentration profile in the leg as a function of the time t during the cycle period, (–).
D	settling leg internal diameter, (m).
D_1	upper and smaller internal diameter of a settling leg with a change in diameter, (m).
D_2	lower and larger internal diameter of a settling leg with a change in diameter, (m).
L	total length of the settling leg, (m).
L_1	length of the upper part, with a smaller internal diameter, of a settling leg with a change in diameter, (m).
L_2	length of the lower part, with a larger internal diameter, of a settling leg with a change in diameter, (m).
S	flux of polymer particles passing a cross section of the settling leg by sedimentation, (m^3 polymer, m^{-2}, s^{-1}).
t	time running during the cycle period T, $t = 0$ at the opening of the leg, and $0 \leq t \leq T$, (s).
T	cycle period: period between two consecutive openings of the same settling leg, (s).
u_D	upward velocity of the interface D between the bottom layer of the thickened slurry at the maximal concentration C_m and the thickening slurry, (m, s^{-1}).
v	average settling velocity of the polymer particles in the emptied fraction of the settling leg, (m, s^{-1}).
v_{sett}	effective settling velocity: average downward velocity by which settling polymer particles enter the emptied fraction of the settling leg, (m, s^{-1}).
x	position on the vertical coordinate axis indicating a position along the leg, $x = 0$ at the interface B, and $x = x_L$ at the bottom of the leg, (m).
η	emptied fraction: fraction of the settling leg's volume, emptied at every opening of the leg, (–).
η_2	of a leg with 2 diameters: the fraction of the total volume taken by the lower part with the larger diameter D_2, (–).

η_f	filled fraction of the settling leg: fraction of the leg volume which is, at the end of the cycle period, filled with polymer packed at the maximal concentration C_m, (–).
ρ	density of the polymer from the reactor, as measured in the lab at room temperature, (kg, m^{-3}).
ρ_{BD}	bulk density of the polymer powder produced in the reactor, as measured in the lab at room temperature, (kg, m^{-3}).

5 Catalyst Activity and Productivity

In the first chapters of this book, the polymerisation reaction itself has not yet been touched upon. This discussion is started now and the concepts of catalyst activity and catalyst productivity are introduced.

5.1 The Polymerisation Reaction

The polymerisation reaction in a slurry loop reactor is started when a certain quantity of catalyst is brought into the reactor liquid at suitable reactor conditions like ethylene concentration, and temperature ...

This catalyst is usually a silica gel particle, called the *carrier*, which presents a large surface due to its porous structure, that has been impregnated with an active chemical[1]. This active chemical is a transition metal brought into a proper state of oxidation and chemical environment, viz. Cr(VI) oxide in the case of a chromium oxide (CrOx) catalyst, $TiCl_4$ in case of a Ziegler–Natta catalyst, or a metallocene or chromocene group for the metallocene and chromocene catalysts.

Part of this active chemical will form *active sites* on the surface of the catalyst particle. An active site is a spatial point on the catalyst surface where the polymerisation proceeds due to the presence of one active atom of the transition metal. A growing polymer chain is attached to an active site.

CrOx catalysts were invented by Hogan and Banks in 1951. They are produced by the impregnation of silica gel particles, SiO_2, with a Cr(III) component, usually chromium acetate in an aqueous solution. To make it active, the impregnated silica is calcined, which means that it is contacted with dry air at high temperatures. This is effected in a fluidized bed reactor, called *activator*, following a strict time-temperature program. At 300 to 400 °C, the Cr(III) species is oxidised to Cr(VI)oxide, CrO_3, and as such chemically bonded to the silica.

A silica surface that has been exposed to moisture, as every silica has before activation, will be covered with silanol groups, ≡Si–OH. These will inhibit the activity, unless further calcination, at temperatures from 550 to 850 °C, removes most of them from the surface. The resulting Cr(VI) atom, bonded to the silica, in an environment relatively free of silanol groups, is the active site.

It is important to note that the silica's role here is more than just offering a support. Without its bond to the silica, CrO_3 is not stable at the required temperatures of calcination, and the catalyst would not exist.

After activation, the catalyst should not meet any trace of moisture, because the bond between silica and CrO_3 is dissolved by water, and the harmful silanol groups would be restored.

---

1 More is said on the silica particle in the next chapter, Section 6.1.

It is generally accepted that the final activation step of this catalyst is the reduction of Cr(VI) to Cr(II) by ethylene upon entering the reactor.
CrOx catalysts are not stereo-specific. They produce a wide molecular weight distribution, which is an advantage or a disadvantage depending on the purpose of the polymer. They cannot be used for polypropylene because they do not yield an isotactic product.

Ziegler–Natta catalysts were named after their inventors in the early 1950s. Initially, Ziegler–Natta catalysts for polyethylene were carried on $TiCl_3$ particles, but, about a decade later, those were replaced by $MgCl_2$ particles, which enhanced the catalyst activity very much. These "salt" particles are rather weak, and are not easily fed by the usual catalyst feeding systems on slurry loop reactors. Ziegler–Natta catalysts supported by silica gel particles are more suitable. The silica gel is calcined, and impregnated with organic complexes providing $MgCl_2$ first, and then $TiCl_4$. The Ti-atom is the active site, while the abundance of Cl^- ions creates the necessary environment, i.e. a high concentration of electrons. This environment is only complete and effective when a cocatalyst, most often triethylaluminium (TEAl), is present in the reactor liquid.
Ziegler–Natta catalysts produce a more narrow molecular weight distribution of the polymer, and are stereo-specific, so that they also allow the production of isotactic polypropylene.

Metallocenes were also known since the early 1950s. But only after the discovery by Kaminsky of methylaluminoxane (MAO), $(-CH_3Al-O-)_n$, and its role as a cocatalyst to metallocenes, in 1977, could they become productive industrial catalysts.
A simple and effective metallocene catalyst is zirconocene, Cp_2ZrCl_2, where Cp stands for a cyclopentadienyl group, i.e. an aromatic ring composed of 5 carbons.
Calcined silica gel particles are impregnated with a high dosis of the MAO cocatalyst, and in a second step the metallocene is added from an organic solution[2]. The metal (Zr) atom is the active site.
Metallocenes are very stereo-specific. They produce the narrowest possible molecular weight distributions. It means that all active sites on the catalyst must show an identical configuration. That is why they are often called *single site catalysts*.

Chromocene catalysts are obtained by the impregnation of a calcined silica gel particle with an organic solution of chromocene, $CrCp_2$[3]. They tend to produce a rather low molecular weight.

Active sites are environments where a bond is established between an ethylene molecule and the active atom, in a way that a polymer chain can be initiated and grow. These environments are created by a distribution of electric loads (electrons), and also

2 W. Kaminsky, *Highly active metallocene catalysts for olefin polymerization*, J. Chem. Soc. Dalton Trans., 1998, pp. 1413–1418.
3 Blom, Dahl, Follestad, Jens, Journal of Molecular Catalysis A: Chemical 138, pp. 97–102, 1999.

by the geometric (steric) configuration of the atoms, since atoms do take space when present.

Detailed reaction mechanisms have been described in the literature. They are sometimes speculative, and their knowledge is not necessary to operate a slurry loop reactor. Still, it is interesting to know that the growing step in the reaction takes place by the insertion of an ethylene molecule in the bond between the active site and the polymer chain. A polymer chain is thus growing from the catalyst particle surface, like a hair on a person's head is growing from the skin. The chain growth is stopped when this bond is broken, e.g. by the insertion of a hydrogen instead of an ethylene molecule, which saturates the double bond, or simply by the thermal motion at the prevailing reactor temperature. When the growth of one polymer chain has stopped, another chain will start to grow at the same active site as long as the site is active, and can be reached by the ethylene molecules.

It is useful to know how much time it takes to form one chain. This time was estimated by Hogan for a CrOx catalyst[4].

The usual concentration of Cr applied to this catalyst by impregnation, is 1.0 *wt*%, which is well in excess of what will cover the silica surface, or will be active.

The activity of one active site[5] is estimated by measuring the activity of a catalyst with a known low concentration of Cr, so low that it can be assumed that all added Cr atoms will be active sites. The activity of one site is then equal to the catalyst activity divided by the number of active sites, where the number of active sites is expressed in moles and is equal to the number of moles of Cr added to the silica.

The concentration of active sites on a catalyst saturated with Cr, is then assumed to be equal to the activity of the catalyst divided by the thus measured activity of one active site.

Another way to determine the concentration of active sites is to poison the catalyst chemically till its activity is completely lost, and to establish the equivalent number of Cr atoms to the amount of poison used.

Hogan and other authors thus arrive at the conclusion that a CrOx catalyst saturated with Cr will contain about $2.5 \cdot 10^{-5}$ kmoles of active Cr sites per *kg* of catalyst.

The time needed to form one polymer chain is then obtained as follows. The number of chains or polymer molecules formed by one *kg* of catalyst per unit time is equal to its activity $\mathbb{A}$ divided by M_n, the number average molecular weight of the polymer. The number of chains formed per unit of time and per active site is then,

$$\frac{\mathbb{A}}{M_n\, 2.5 \cdot 10^{-5}},$$

4 Hogan J. P., *Ethylene Polymerization Catalysis over Chromium Oxide*, Journal of Polymer Science, volume 8, pp. 2637–2652, 1970.

5 The term *activity* has not yet been defined here, but can be understood as the quantity of polymer produced per unit of time at one *kg* of catalyst, or at one active site on the catalyst.

and the average time t_1 needed to form one chain is,

$$t_1 = \left(\frac{\mathbb{A}}{M_n\, 2.5 \cdot 10^{-5}} \right)^{-1}.$$

The calculated time, with activities around $4,000\ kg\ kg^{-1}\ h^{-1}$ and M_n around 10,000 is of the order of magnitude of one second. This is a very comforting piece of information. It means that the time to form one polymer molecule is much smaller than the average catalyst residence time, which is of the order of one hour. The catalyst residence time will thus not affect the quality of the polymer. The product will be reproducible at all production rates, and test runs at low production rates will yield representative polymers.

A Note on Health and Environmental Hazards from CrOx Catalysts
Trivalent chromium components are known as allergenic, whilst hexavalent chromium compounds are carcinogenic. Both affect the respiratory tract. The hazards are well known from the use of chromic acid baths for chromium plating of metal objects, and the welding of stainless steel.
Even if Cr(III) compounds are less toxic than Cr(VI) compounds, – the occupational exposure limits for Cr(III) are 10 times higher than for Cr(VI) –, it is recommended not to discriminate between the two hazards, and to avoid all exposure to CrOx catalysts, before as well as after activation. This requires a substantial effort in the providing of adequate operating procedures, and especially in the lay-out of the equipment in the catalyst activation and feeding sections of the plant.

The fact that moisture is dissolving the bond between the CrO_3 species and the silica, and desactivates the CrOx catalyst, must not lead to the treacherous misunderstanding that water would reduce hexavalent chromium to its trivalent state. Cr(VI) components are highly soluble and stable in water. When treating spills or wastes, a reducing agent like sodium sulphite in an acid environment must be applied.

The CrOx catalyst is, like all other catalysts in the slurry loop process, not recovered from the polymer, and so is left in the produced polyethylene. Fortunately, it is reduced from the hexavalent to the bivalent state in the reactor, due to a reaction with ethylene. Anti-oxidant stabilizers added to the polymer during the subsequent extrusion step in the process, are also strong reducing agents and would reduce the hexavalent chromium, if still present.

Though chromium is classified as a heavy metal, its concentration in HDPE, of the order of some 5 *ppm*, is lower than in many other substances found in the environment. □

5.2 Polymerisation Rate Equation

Reaction rate equations have been conceived for homogeneous reactions. Their application to slurry loop reactors is not straightforward. The reactor's volume is not homogeneous, being shared by a solid phase, the catalyst encapsulated by growing polymer, and a liquid phase. And the ethylene polymerisation reaction is not homogeneous either, proceeding at the interface of a solid and the reactor liquid.

The polymerisation reaction rate is usually represented by a *first order rate equation* for ethylene,

$$r = k(T)y_2 y_a ,$$

where r is the *reaction rate* (kmoles of ethylene polymerised per unit time, per unit volume of reactor liquid); $k(T)$ is a *reaction rate constant* which is not dimensionless; y_2 is the *ethylene concentration* (moles of ethylene per unit volume of reactor liquid); and y_a is the *concentration of active sites* (number of active sites present per unit volume of reactor liquid).

The obstacle of the inhomogeneous volume is avoided by referring to the volume of the liquid phase. And so is the obstacle of the inhomogeneous reaction by introducing the concentration of active sites related to the liquid volume.

The active sites are not all identical. The reaction rate is necessarily an average value over the different active sites.

The rate equation is a first order equation according to the power equal to 1 of the ethylene concentration y_2. A first order is often approximative when applied to represent a complex reaction mechanism by an empirical rate equation. But a first order may also be correct when diffusion of the reactant to the active site is the rate determining step. The latter is probably the case for the polymerisation of ethylene.

The rate constant $k(T)$ will in general depend on the temperature by an Arrhenius function $k = Z \exp E_a/RT$. Besides temperature, all other differences between reactor systems will be absorbed in the rate constant, which, arguably, is all but a constant.

The concentration of active sites may be expressed by means of n_a, the *number of active sites per unit weight of catalyst.* When that number is incorporated in the rate constant, the rate equation may be like,

$$r = k(T)y_2 n_a y_{cat} = k(T)n_a y_2 y_{cat} = k_2'(T)y_2 y_{cat} ,$$

where y_{cat} is the *catalyst concentration* (*kg* of catalyst per unit volume of reactor liquid).

Finally, the reaction rate r may be expressed as the number of kilogrammes of ethylene polymerised per unit time and per unit weight of catalyst present in the reactor. The rate equation will then take its simplest form when transformed from,

$$r = k_2'(T)y_2 y_{cat} ,$$
$$r(\mathrm{MW})/y_{cat} = k_2'(T)(\mathrm{MW})y_2 ,$$

to,

$$r_2 = k_2(T)y_2 ,$$

where r_2 is the *reaction rate* (*kg* of ethylene per second, per *kg* of catalyst), and $k_2(T)$ is a *reaction rate constant.* This rate equation is most close to the simple fact that a catalyst submersed in a liquid containing ethylene is polymerizing ethylene, and is most practical when applied to the mass of catalyst present in the reactor.

The rate equation suggests that the reaction rate be constant in time. This is far from true because the rate of reaction shows a tendency to diminish with time. Each catalyst particle, producing its own polyethylene, is encapsulating itself with this polymer, and hampering the diffusion to its active sites more and more.
That is why the reaction rate is in practice represented by an exponentially declining function of the catalyst lifetime. By *lifetime t* is meant the time that a catalyst particle, present in the reactor, has already spent there since it was injected into that reactor. A practical rate equation thus looks like,

$$r_2 = k_2(T)\mathrm{e}^{-at}y_2 \,. \tag{5.1}$$

5.3 Catalyst Activity

The *catalyst activity* is, in general, defined as the quantity of polymer produced by one *kg* of catalyst per unit time. This definition should be more specific to be useful.

By catalyst activity is often meant a rather gross average value for all catalyst leaving the reactor, calculated as the *total reactor production rate, P*, (*kg*, h^{-1}), divided by the *catalyst feed*, F_{cat}, (*kg*, h^{-1}), and the *average catalyst residence time*, τ (*h*),

$$\text{activity} = \frac{P}{F_{\text{cat}}\tau} \,.$$

Here two more narrow definitions are used.

The *catalyst activity*, $\overline{A}(t)$, is defined as the rate of polymer produced by one *kg* of catalyst per unit time at a certain lifetime t. It is shown as an average value because a quantity of catalyst contains different particles which have active sites of different natures and activities, and the value of $\overline{A}(t)$ is an average over those particles and active sites at lifetime t.
The catalyst activity, as such defined, is represented by the empirical expression,

$$\overline{A}(t, Rx) = b\mathrm{e}^{-at} \,, \tag{5.2}$$

which will be elucidated later, and where Rx stands for a *set of reactor conditions*. It is clear that, at a given set of reactor conditions Rx, the catalyst activity equation (5.2) is nothing else than the polymerisation rate equation (5.1).
The function $\overline{A}(t, Rx)$ plotted as a function of t is called the *catalyst activity profile*.

When the catalyst is used in a particular reactor, the *catalyst reactor activity*, $\mathbb{A}$, is defined as the rate of polymer produced by one *kg* of catalyst present in that reactor per unit time. This activity is called *reactor* activity because it is the combined result of the activities of different fractions of catalyst present in the reactor at the same time, though at different lifetimes t, and will, besides on the catalyst itself, also depend on the catalyst lifetime distribution in the reactor.

The *catalyst lifetime distribution* $E_1(t)$ is a function of the catalyst lifetime t such that $E_1(t)\mathrm{d}t$ is equal to the weight fraction of the catalyst with a lifetime between t and $(t + \mathrm{d}t)$.

The fraction of 1 *kg* of catalyst present in the reactor, that owns a lifetime t is equal to $E_1(t, Rx)\mathrm{d}t$. The quantity of polymer produced in the reactor per unit of time by this fraction of the catalyst with lifetime t is,

$$\mathrm{d}\mathbb{A} = E_1(t, Rx)\overline{\mathcal{A}}(t, Rx)\mathrm{d}t\,.$$

The catalyst reactor activity $\mathbb{A}$ is the quantity of polymer produced per unit of time by 1 *kg* of catalyst present in the reactor, and is obtained by summing over all values of lifetime,

$$\mathbb{A} = \int_{t=0}^{\infty} E_1(t, Rx)\overline{\mathcal{A}}(t, Rx)\mathrm{d}t\,. \tag{5.3}$$

5.4 Catalyst Reactor Productivity

The catalyst activity is a property of the catalyst during its lifetime inside the reactor. The catalyst reactor productivity, on the contrary, is a property of the catalyst that has left the reactor.

The *catalyst reactor productivity*, $\mathbb{P}$, (*kg* polymer per *kg* of catalyst), is the total quantity of polymer produced by one *kg* of catalyst that has left the reactor. It is called *reactor* productivity because it is the combined result of the productivities of different fractions of catalyst with different residence times θ in the reactor, and will also depend on the catalyst residence time distribution. The *catalyst residence time*, θ, is the total time that the catalyst has spent in the reactor at the moment when it leaves the reactor.

The *catalyst residence time distribution*, $E_2(\theta)$, is a function of the catalyst residence time θ such that $E_2(\theta)\mathrm{d}\theta$ is equal to the weight fraction of the catalyst with a residence time between θ and $(\theta + \mathrm{d}\theta)$.

The fraction of 1 *kg* of the catalyst, leaving the reactor, that owns a residence time θ is equal to $E_2(\theta, Rx)\mathrm{d}\theta$.

When the catalyst leaves the reactor at a certain residence time θ, its lifetime t has run from zero to θ, and the catalyst has been active according to its activity profile $\overline{\mathcal{A}}(t, Rx)$. The quantity of polymer produced by the fraction of catalyst with residence time θ is,

$$\mathrm{d}\mathbb{P} = E_2(\theta, Rx)\mathrm{d}\theta \int_{t=0}^{\theta} \overline{\mathcal{A}}(t, Rx)\mathrm{d}t\,.$$

The catalyst reactor productivity is the total quantity of polymer produced by 1 *kg* of catalyst, and is obtained by summing over all fractions of catalyst with all possible residence times,

$$\mathbb{P} = \int_{\theta=0}^{\infty} E_2(\theta, Rx) \left[\int_{t=0}^{\theta} \overline{\mathcal{A}}(t, Rx) \mathrm{d}t \right] \mathrm{d}\theta \,. \tag{5.4}$$

5.5 Catalyst Reactor Productivity in a Batch Reactor

Catalyst productivities are tested in small batch reactors before use on commercial scale.

To this end, a quantity of catalyst m_{cat} is brought into the reactor at the required reactor conditions and the polymerisation is started by the feeding of ethylene. The pressure in the reactor is kept constant by modulation of the ethylene feed flow. The quantity of ethylene consumed to maintain the constant pressure is equal to the ethylene that has been polymerised in the reactor. In a batch reactor all catalyst owns the same history and the same lifetime. When the reaction is stopped by depressurising the reactor, the total duration of the polymerisation experiment is the lifetime reached by all catalyst and is equal to the common residence time.

The catalyst activity profile is assumed to be,

$$\overline{\mathcal{A}}(t, Rx) = b\mathrm{e}^{-at} \,.$$

There is a uniform catalyst lifetime distribution in the reactor, all catalyst having the same lifetime t, which is equal to the time elapsed since the start of the experiment. So the catalyst reactor activity at the time t is equal to,

$$\mathbb{A}(t, Rx) = \overline{\mathcal{A}}(t, Rx) = b\mathrm{e}^{-at} \,,$$

and is declining with time.

There is a uniform catalyst residence time distribution in the reactor. When the reaction is stopped, all catalyst has the same residence time, which is equal to the total duration of the experiment. The catalyst reactor productivity in a batch reactor is thus equal to,

$$\mathbb{P} = \int_{t=0}^{\theta} \overline{\mathcal{A}}(t, Rx) \mathrm{d}t = \int_{t=0}^{\theta} b\mathrm{e}^{-at} \mathrm{d}t \,,$$

or,

$$\mathbb{P} = \frac{b}{a}(1 - \mathrm{e}^{-a\theta}) \,. \tag{5.5}$$

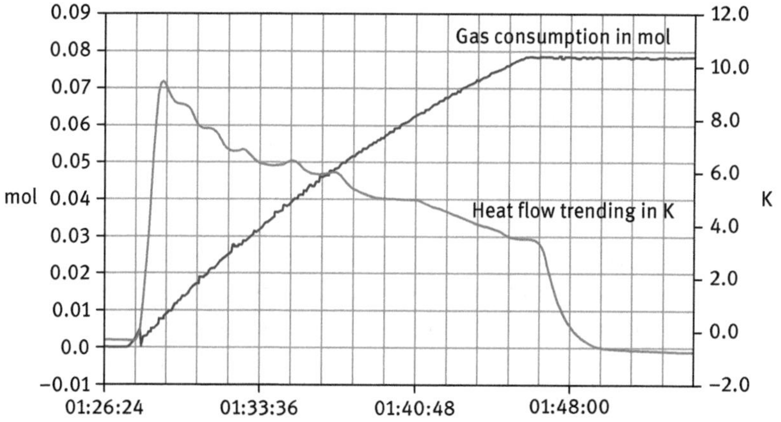

Fig. 5.1. Batch polymerisation test of ethylene (Visentin and Wagner, 2008).

A batch polymerisation of ethylene by a metallocene catalyst in a 50 ml reactor has been presented by Visentin and Wagner[6], Figure 5.1.

The catalyst activity is measured by the reaction heat, which is proportional to the temperature difference $(T_r - T_j)$ between the reactor and the cooling liquid in the reactor cooling jacket.

The catalyst reactor productivity is measured as the accumulated ethylene consumption.

The catalyst reactor productivity curve (which is the most precise curve) can be quantified (in moles of ethylene and seconds) as,

$$m_{\text{cat}}\mathbb{P} = \frac{m_{\text{cat}} b}{a}(1 - \mathrm{e}^{-a\theta}) = 0.1281(1 - \mathrm{e}^{-3.1881\theta})\,.$$

The obtained value of $a = 3.1881\,s^{-1}$ allows to fit the activity curve very well as,

$$m_{\text{cat}}\mathbb{A}(t, Rx) = m_{\text{cat}} b\mathrm{e}^{-at} = 9.1\mathrm{e}^{-3.1881\theta}\,,$$

which confirms the expressions (5.2) and (5.4) presented here.

In larger, so called bench scale, batch reactors, with volumes of 2 and more litres, the recorded activity and productivity curves often deviate from what is expected, probably due to diffusion phenomena. However, curves derived from full scale reactors do match the presented empirical expressions.

6 Fabio Visentin, Matthias Wagner, *Polymerization of ethylene and propylene …*, Thermal Analysis UserCom, 27, Mettler-Toledo AG, 2008.

Notations

a	constant in the expressions for the catalyst activity profile and catalyst productivity, (s^{-1}).
$\mathbb{A}$	catalyst reactor activity: quantity of polymer produced by a quantity of catalyst present in the reactor per unit time, (*kg* polymer/*kg* catalyst, s).
$\overline{A}(t, Rx)$	catalyst activity: quantity of polymer produced by one *kg* of catalyst per unit time at a certain lifetime t and a set of conditions in the reactor, (*kg* polymer/*kg* catalyst, s).
b	constant in the expressions for the catalyst activity profile and catalyst productivity, (kg, kg^{-1}, s^{-1}).
$E_1(t)$	catalyst lifetime distribution: a function of the catalyst lifetime t such that $E_1(t)\mathrm{d}t$ is equal to the weight fraction of the catalyst with a lifetime between t and $(t + \mathrm{d}t)$, (s^{-1}).
$E_2(\theta)$	catalyst residence time distribution: a function of the catalyst residence time θ such that $E_2(\theta)\mathrm{d}\theta$ is equal to the weight fraction of the catalyst with a residence time between θ and $(\theta + \mathrm{d}\theta)$, ($s^{-1}$).
E_a	energy of activation in the Arrhenius equation, ($J, kmole^{-1}$).
F_cat	catalyst feed to the reactor, (kg, s^{-1}).
$k(T), k_2'(T), k_2(T)$	reaction rate constants.
m_cat	mass of catalyst present in the reactor, (*kg*).
M_n	number average molecular weight of the polymer, ($kg, kmole^{-1}$).
MW	molecular weight of ethylene, ($kg, kmole^{-1}$).
n_a	number of active sites per unit weight of catalyst, (kg^{-1}).
P	polymer production rate: total quantity of polymer produced in the reactor per unit of time, (*kg* polymer, s^{-1}).
$\mathbb{P}$	catalyst reactor productivity: total quantity of polymer produced by one *kg* of catalyst that has left the reactor, (*kg* polymer/*kg* catalyst).
r	reaction rate: quantity of ethylene polymerised per unit time, per unit volume of reactor liquid, ($kmole, s^{-1}, m^{-3}$).
r_2	reaction rate: quantity of ethylene polymerised per unit time, per *kg* of catalyst, (kg, s^{-1}, kg^{-1}).
R	the gas constant, ($J, kmole^{-1}, K^{-1}$).
Rx	symbol representing all of the reactor conditions, including temperature, concentrations, and flows.
t	catalyst lifetime: the time interval between the moment when a catalyst particle is fed to the reactor and any moment at which the reactor system is considered, (s).
t_1	time needed to form one polymer chain on an active site, (s).

T	absolute temperature, (K).
y_2	ethylene concentration, in moles of ethylene per unit volume of reactor liquid, ($mole, m^{-3}$).
y_a	concentration of active sites, as number of active sites present per unit volume of reactor liquid, (m^{-3}).
y_{cat}	catalyst concentration, as kg of catalyst per unit volume of reactor liquid, (kg, m^{-3}).
Z	integration constant in the Arrhenius equation.
θ	catalyst residence time: the total time that the catalyst has spent in the reactor at the moment when it leaves the reactor, (s)
τ	average catalyst residence time, (s).

6 The 1/1 Hypothesis

Before further developing thoughts on catalyst activity, productivity and residence time in Chapter 7, it is necessary to present the *one catalyst particle, one polymer particle assumption*, or *1/1 hypothesis*. This assumption is stating that each catalyst particle, that is fed to the reactor, is leaving it as one polymer particle, containing all of the polymer it has produced. It is underlying all expressions of catalyst activity and productivity.

6.1 Silica Particles

Most catalysts used for polymerisation in slurry loop reactors are supported by silica gel particles, of a well defined porosity and particle size.

Silica gel, SiO_2, is an amorphous and highly porous substance, not to be confused with quartz, the crystalline material of the same chemical composition, but totally different structure. The roots of the silica gel production are found with soap manufacturing, one of the oldest branches of the chemical industry[1]. Silica gel is produced by reacting sodium silicate (waterglass) with dilute sulphuric acid. The reaction product is a *silica hydrosol*, i.e. a suspension of oligomers of silicic acid molecules, $Si(OH)_4$, from which a number of water molecules have been extracted by a condensation reaction. The oligomers thus consist of siloxane chains, –O–Si–O–, completed with hydroxyl groups, –OH, on the Si atoms.

The hydrosol is washed with water to remove sodium sulphate, the other reaction product.

The condensation reaction continues, and the hydrosol is turned into a *silica hydrogel*, an amorphous threedimensional structure, due to the crosslinking of the oligomers. The art of making a silica gel with a well defined pore structure lies with the manipulation of this initial hydrogel. It can be dried immediately or aged. In the latter *aging process* the hydrogel is kept wet, and parameters like time, temperature and acidity are carefully controlled, before the final drying.

The required particle size is obtained by grinding and sieving. Particles thus produced are called *granular* particles.

An alternative route is the spray-drying of a suspension of the hydrogel. The so-called *micro-spheroidal* particles thus obtained, show a nice spherical particle morphology.

The active sites of chromium oxide (CrOx) catalysts can be modified by adding other elements like titanium, aluminium, or fluorine to the silica. This can be effected by impregnation, but also by incorporating these elements in the gel structure, which

1 See A. E. Musson, *Enterprise in Soap and Chemicals; Joseph Crosfield & Sons Ltd 1815–1965*, Manchester University Press, 1965. Another unit operation due to the soapmaking industry is hydrogenation.

is then called a *cogel*[2]. Silica-titania and silica-alumina cogels are obtained by adding Ti- or Al-compounds to the initial reacting sodium silicate solution. Such modified catalysts are claimed to produce better polymer properties, due to a broader molecular weight distribution, but if this broadening just means the addition of a low molecular weight tail, there is not necessarily a great advantage to be found!

Another modification of CrOx catalysts is obtained by using organic solvents instead of water, either in the washing of the hydrosol, or in the impregnation with a Cr(III) component. Higher pore volumes may be obtained, but the manufacturing and environmental costs are increased.

A silica gel catalyst support, as well as the thus supported catalyst, is commercially described by average particle size, surface area, and pore volume.

The *average particle size*, APS, (μm), is measured by laser diffraction techniques and lies between 25 and 150 μm. The particle size distribution should be as narrow as possible. Tails of fines should be avoided. Finer particles are usually more active.

The *surface area*, SA, (m^2, g^{-1}), is all area presented by the particle to the surroundings, both on the outer surface and on the walls of the pores. It is measured by the BET nitrogen sorption method and varies between 200 and 500 m^2g^{-1}.

The *pore volume*, PV, (cm^3, g^{-1}), is the total empty volume constituted by the pores. It is measured by water titration or BET nitrogen sorption and varies between 1 and 3 cm^3g^{-1}. The pores are classified as macropores, mesopores and micropores, according to their pore diameter being larger than 10 nm, between 10 and 1 nm, or less than 1 nm respectively.

It is difficult to relate the polymer properties to the catalyst particle properties. The *average pore diameter*, APD, (Å), seems to be the most determining parameter for the melt index (or average molecular weight) level of the polymer, though it is seldom specified. It is calculated as the diameter of a hypothetical cylinder with volume PV and side surface SA, and equals APD $= 4 \cdot 10^4 \cdot$ PV/SA. The larger the APD, the higher the produced melt index, or the lower the average molecular weight.

More detailed information, often helpful to solve problems, is provided by the pore diameter distribution function.

The silica particles are quite fragile, which they should be, due to their high porosity. This high porosity is illustrated by their low density and bulk density. The density of the solid silica "walls" between the pores is estimated at 2250 $kg\ m^{-3}$. The volume of one kg of porous silica is thus equal to (1/2250 + PV/1000), and the density of a porous silica is equal to,

$$\rho_{cat} = \left(\frac{1}{2250} + \frac{PV}{1000}\right)^{-1},$$

or between 300 and 700 $kg\ m^{-3}$.

2 C. E. Marsden, *Advances in Supported Chromium Catalysts*, Plastics, Rubber and Composites Processing and Applications, volume 21, pp. 193–200, 1994.

The bulk density of a powder of such particles, assuming them to be spherical, and assuming ideal packing, would be $0.5236(1/2250 + PV/1000)^{-1}$.
An empirical correlation for the bulk density of granular silica gel powders, and thus supported catalyst powders yields,

$$\rho_{\text{cat, BD}} = 0.5635\left(\frac{1}{2250} + \frac{\text{PV}}{1000}\right)^{-1},$$

or from 160 to 400 $kg\ m^{-3}$.

The external surface of the particle is negligible compared to the internal pore surface[3], and so the activity of the catalyst, generated by the active sites on its surface, must take place inside the pores.

6.2 Catalyst Particle Fragmentation and Growth

The diffusion of the reagents inside the pores would soon be hemmed in by the produced polymer, and the reaction would cease, if not, as is generally accepted in the literature, the silica particles would break up into fragments during the first few minutes of the reaction. The size of those fragments is of the order of 1 μm. By this fragmentation the access of the reagents to the active places is maintained. More fragile catalyst particles thus exhibit a higher activity. This necessary friability is promoted by high pore volumes, but also the strength of the silica walls between the pores, which depends on the manufacturing process of the silica, is important.

In spite of this thorough fragmentation, the catalyst particle continues its life in the reactor as one single particle, because its fragments are cemented together by the generated polymer. This polymer makes the catalyst particle grow into a single polymer particle of a size some 30 times larger, and drives the dispersed active sites to the outside, and thus accessible to the reagents. Still, a number of active sites will slowly be closed in by the polymer, and the catalyst particle will lose of its activity.
Hence the *one catalyst particle, one polymer particle assumption* or briefly called the *1/1 hypothesis*, which is stating that one catalyst particle, after entering the reactor, will not split nor agglomerate with other particles, and leave the reactor as one polymer particle, containing all of the polymer it has produced.
The 1/1 hypothesis also includes that, since a single catalyst particle can be traced on its way through the reactor, it also keeps its original properties. E.g., high catalyst activities often go together with small catalyst particle sizes. It is assumed that these small particles keep their high activity, and all of their other properties that are due

3 This is illustrated by a small example. The external surface area of a 100 μm perfect sphere is calculated at $\pi \cdot 10^{-8}\ m^2$. If a spherical porous particle of that size owns a PV of 2.0, its density is 409 $kg\ m^{-3}$ and its weight is $68\pi \cdot 10^{-9}\ g$. The external surface area of the particle is thus $0.15\ m^2g^{-1}$, to be compared to the total SA of several hundreds of m^2g^{-1}.

to the original particle size distribution, even if, by fragmentation, nothing is left of the original particle size. In that sense, it is stated that the catalyst particle size, as a property of a catalyst particle, is not changed in the reactor.

Another important feature of this process is that the shape of the original catalyst particle is conserved in the polymer particle. The silica carrier particle is growing uniformly in all directions and so the resulting polymer particle becomes a larger size replica of the original. This fact is sometimes referred to as the *replica phenomenon.* It means that the morphology of the polymer particles can be steered by the morphology of the silica, which is important to their settling properties, or to the pumping characteristics of the slurry. Spheroidal particles, e.g., show certain advantages over granular particles.

Together with the replica phenomenon, it must be considered that small catalyst particle sizes will produce small polymer particles, which may affect the properties of the slurry. A too fine polymer may be impossible to be pumped as a slurry. Average particle sizes of catalysts for slurry loop reactors are usually at least 25 μm, and tails of fines to the particle size distribution should be avoided.

A surprising and interesting illustration of all this is shown in Figure 6.1. It shows a number of 10 $Å$ molecular sieve granules taken from an ethylene dryer bed that was plugged by the polymerisation of ethylene inside the granules by repeated insufficient cooling after regeneration. Probably some transition metals, present in the extrudate granules, had acted as a catalyst to the ethylene, while the molecular sieve had served as the catalyst support. Several extrudate particles were sampled which had reached different stages of polymerisation, showing a kind of slow motion representation of what happens to a catalyst particle in the reactor. The original extrudate is seen on the

Fig. 6.1. Fragmentation and growth of a 10 $Å$ molecular sieve extrudate granule in an ethylene dryer by polymerisation of ethylene in its pores. With thanks to Franco Zori.

left. The second particle from the left is already cracked by the beginning polymerisation. The further stages, from left to right, are self-explaining and show the fragmentation, the growth, the coherence of the original particle, and the conservation of the original morphology.

6.3 Catalyst Particle Productivity

The catalyst reactor productivity $\mathbb{P}$ is the sum of the polymer productions from different fractions of catalyst leaving the reactor after different residence times. The 1/1 hypothesis allows to consider one single catalyst particle, or, in reality, a number of catalyst particles that have followed the same path with respect to concentrations and lifetime, and to define the productivity of that particle. This concept will be useful when discussing the production of polymer in two reactors in series.

The *catalyst particle productivity*, $\mathcal{P}$, is defined as the quantity of polymer amassed by the catalyst particle on its own surface during the course of its life in the reactor, or, the weight of polymer found in one particle leaving the reactor.
In realistic terms it is calculated as the productivity of one *kg* of catalyst particles that have followed the same path in the reactor,

$$\mathcal{P} = \int_{t=0}^{\theta} \overline{A}(t, Rx)\mathrm{d}t \,. \tag{6.1}$$

The particle productivity, $\mathcal{P}$, is the result of the history of a single particle on its path of concentrations met in the reactor. Its polymer product is only found on that particle. It is a function of the residence time θ, and is an average value over all catalyst particles that have followed the same path.

The catalyst particle productivity is mathematically equal to the catalyst reactor productivity obtained in a batch reactor, as described in Section 5.5. Which is not surprising, since in a well mixed batch reactor all particles have followed the same path, and the reactor productivity, common to all particles, must coincide with the particle productivity. But that only holds for a batch reactor. Another view on this mathematical identity is to consider each catalyst particle as a small batch reactor, where, according to the 1/1 hypothesis, all polymer formed stays together and shares the same path.

6.4 Balance for the Number of Particles

Another consequence of the 1/1 hypothesis is that the number of particles comprised in one kilogramme of catalyst fed to the reactor is related to the number of polymer particles found in one kilogramme of produced polymer. This relation allows to verify the 1/1 hypothesis itself.

The catalyst fed to the reactor owns a certain *particle size distribution*, $w_0(x)$, which is a function of the catalyst particle size x such that $w_0(x)\mathrm{d}x$ is equal to the weight fraction of the catalyst feed with a particle size between x and $(x + \mathrm{d}x)$.
The particle size distribution of the catalyst feed can also be based on the number of particles instead of the weight. Let then $n_0(x)$ be that *particle size distribution based on numbers*, or, a function of x such that $n_0(x)\mathrm{d}x$ is equal to the fraction of the number of particles in the catalyst feed with a particle size between x and $(x + \mathrm{d}x)$.

Assuming that all catalyst particles are spherical, $n_0(x)$ can be derived from $w_0(x)$ as,

$$n_0(x)\mathrm{d}x = \frac{w_0(x)\mathrm{d}x\left(\rho_{\mathrm{cat}}\frac{\pi x^3}{6}\right)^{-1}}{\int_{x=0}^{\infty} w_0(x)\mathrm{d}x\left(\rho_{\mathrm{cat}}\frac{\pi x^3}{6}\right)^{-1}},$$

where ρ_{cat} is the *particle density of the catalyst, (kg, m^{-3})*, or the density of the porous silica that forms the particle, including the void of the pores.
The denominator is the *total number of catalyst particles comprised in* 1 *kg of catalyst feed*, and will be represented as α_0.
Hence,

$$n_0(x)\mathrm{d}x = w_0(x)\mathrm{d}x\left(\rho_{\mathrm{cat}}\frac{\pi x^3}{6}\right)^{-1}\alpha_0^{-1}.$$

By analogy, the *particle size distributions* $w_2(x)$ and $n_2(x)$ and the *total number of particles per kg*, α_2, *for the catalyst in the product stream leaving the reactor*, can be defined, with,

$$n_2(x)\mathrm{d}x = w_2(x)\mathrm{d}x\left(\rho_{\mathrm{cat}}\frac{\pi x^3}{6}\right)^{-1}\alpha_2^{-1}.$$

This catalyst particle size x in the product stream is rather artificial. The volume of polymer adhering to the catalyst is not considered. Neither is, as explained in the previous section, the fact that the catalyst particle has been fragmented by the force of the polymerisation reaction. In other words, it simply means the particle size of the catalyst particle as it was when fed to the reactor.

Under steady state conditions the flow of catalyst leaving the reactor in the product stream will be equal to the *catalyst feed*, F_{cat}, *(kg, s^{-1})*.
Also under steady state conditions, and the 1/1 hypothesis, the number of catalyst particles present in the reactor is constant.
Hence, the rate of catalyst particles entering the reactor will be equal to the rate of catalyst particles leaving the reactor,

$$F_{\mathrm{cat}}\alpha_0 = F_{\mathrm{cat}}\alpha_2,$$

and,

$$\alpha_0 = \alpha_2.$$

Not only the total number of catalyst particles, but also the number of catalyst particles of one particular particle size will be constant in the reactor, at steady state, or,

$$F_{\text{cat}}\alpha_0 n_0(x)\mathrm{d}x = F_{\text{cat}}\alpha_2 n_2(x)\mathrm{d}x\,.$$

Whence it follows that $n_0(x) = n_2(x)$ and further that $w_0(x) = w_2(x)$.

According to the 1/1 hypothesis, the number of *polymer* particles per *kg* of catalyst in the product stream is equal to α_2, the number of *catalyst* particles per *kg* of catalyst in the product stream.

The weight of polymer produced per *kg* of catalyst is the catalyst reactor productivity $\mathbb{P}$.

So the *number of polymer particles per kg of polymer in the product stream*, $\alpha_{2,\text{pol}}$, will be $\mathbb{P}$ times smaller than the number of polymer particles per *kg* of catalyst in the product stream, or,

$$\alpha_{2,\text{pol}} = \frac{\alpha_2}{\mathbb{P}} = \frac{\alpha_0}{\mathbb{P}}\,. \tag{6.2}$$

Hence the conclusion that the number of polymer particles per kilogramme of polymer in the product stream is equal to the number of catalyst particles per kilogramme of catalyst in the catalyst feed, divided by the catalyst reactor productivity.

This proposition is used to confirm the 1/1 hypothesis from experimental data.

6.5 Verification of the 1/1 Hypothesis

To verify the 1/1 hypothesis making use of equation (6.2), the numbers of particles in one kilogramme of catalyst and polymer should be "counted ".

The number of particles per kilogramme catalyst and polymer are derived from their particle size distributions (PSD). Those are obtained by a laser diffraction analysis for the catalyst, and by a sieve analysis for the polymer powder, provided that a sufficient number of sieves is used.

A convenient way to represent a PSD is to plot the *cumulative PSD as the weight fraction of the powder with a particle size smaller than x*, $W(x)$, versus the logarithm of the particle size $\log x$. An example is shown in Figure 6.2 for an activated CrOx catalyst, and in Figure 6.3 for the polymer powder produced by that same catalyst.

In that type of chart, the PSD is approximated fairly well by connecting the measured points by straight lines. So the following calculations are based on the assumption that the cumulative PSD $W(\log x)$ is linear between two measured particle sizes x_1 and x_2.

The cumulative PSD for either a catalyst or polymer powder in a particle size interval $x_1 < x < x_2$ is thus,

$$W(x) = W(x_1) + \frac{W(x_2) - W(x_1)}{\log(x_2/x_1)}(\log x - \log x_1)\,.$$

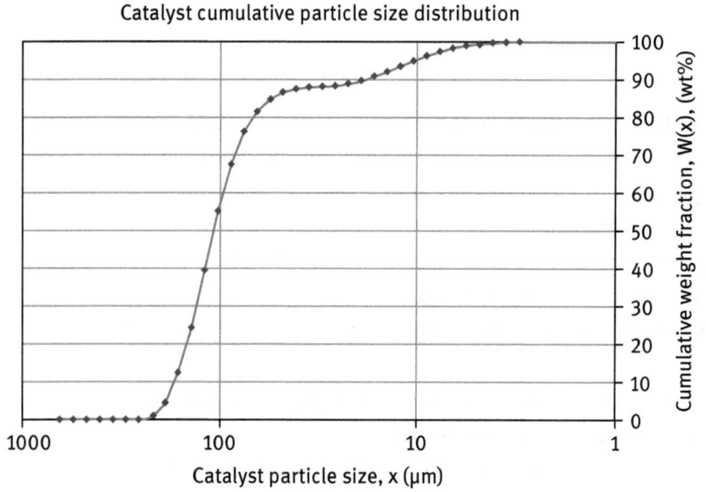

Fig. 6.2. Cumulative Particle Size Distribution for a CrOx Catalyst.

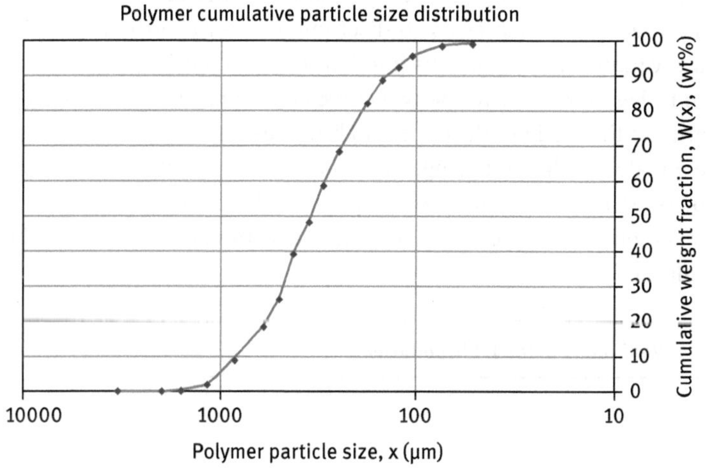

Fig. 6.3. Cumulative Particle Size Distribution for the Polymer produced by the same CrOx Catalyst as in Figure 6.2.

The particle size distribution $w(x)$ in the same interval is,

$$w(x) = \frac{\mathrm{d}W(x)}{\mathrm{d}x} = \frac{W(x_2) - W(x_1)}{\log(x_2/x_1)} \frac{1}{x} \,.$$

The *number of particles comprised in the fraction of the powder with particle size* x *between* x_1 *and* x_2 is,

$$\alpha_{12} = \int_{x_1}^{x_2} w(x) \left(\rho \frac{\pi x^3}{6} \right)^{-1} \mathrm{d}x = \frac{W(x_2) - W(x_1)}{\log(x_2/x_1)} \frac{6}{\rho\pi} \int_{x_1}^{x_2} \frac{\mathrm{d}x}{x^4} \,,$$

or,

$$\alpha_{12} = \frac{W(x_2) - W(x_1)}{\log(x_2/x_1)} \frac{2}{\rho\pi} \left(x_1^{-3} - x_2^{-3}\right) . \tag{6.3}$$

To verify the 1/1 hypothesis, the number of particles α_{12} is calculated for all fractions of particle sizes in the PSD and totalled, for catalyst and polymer powder. The total number of catalyst particles, divided by the catalyst reactor productivity should equal the total number of polymer particles.

However, at smaller particle sizes, the denominator in equation (6.3) tends to zero, and the numbers of particles contributing to the total number, by those fractions, tend to infinity. The error thus introduced by the fractions of small particles may forbid a true comparison.

Therefore, instead of reporting just the two sums, the numbers of particles are totalled for the different weight fractions of the cumulative PSD's of the catalyst and the polymer, and the results (divided by the productivity for the catalyst) are plotted, starting from the largest size and going downwards to smaller and smaller sizes. When the 1/1 hypothesis holds, the two curves should meet at 100% weight fraction at the total number of polymer particles. In reality, the curves will often intersect close to that point, and some interpretation is necessary to discriminate between the trend of a curve and the run-away induced by erroneous fractions of small particles.

An example is given in Figure 6.4 where the two curves nicely meet at 100% weight fraction, thus confirming the 1/1 hypothesis. Though this example is not an exception, the results are not always that flattering. Often a fine particle tail in the catalyst PSD, as measured by laser diffraction, is blowing up the total number of particles. Still, this verification is convincing enough to state that, also from full scale data, the 1/1 hypothesis is confirmed.

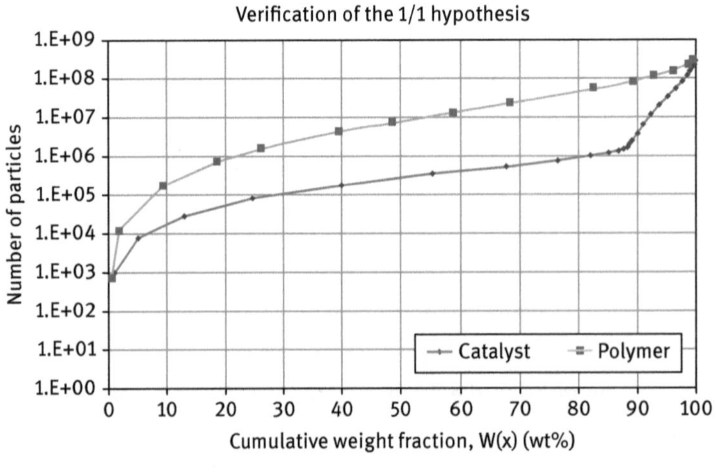

Fig. 6.4. Verification of the 1/1 Hypothesis for the same CrOx Catalyst as in Figures 6.2 and 6.3.

Notations

$\overline{\mathcal{A}}(t, Rx)$	catalyst activity: quantity of polymer produced by one *kg* of catalyst, per unit time, at a certain lifetime *t* in the reactor, (*kg* polymer/*kg* catalyst, *s*).
APD	average pore diameter of a silica gel catalyst support particle, or silica gel supported catalyst particle, (*Å*).
APS	average particle size of a silica gel catalyst support particle, or silica gel supported catalyst particle, (μm).
F_{cat}	catalyst feed to the reactor, (kg, s^{-1}).
$n_0(x)$	particle size distribution of the catalyst in the catalyst feed, based on the number of particles: a function of x such that $n_0(x)\mathrm{d}x$ is equal to the fraction of the number of particles in the catalyst feed with a particle size between x and $(x + \mathrm{d}x)$, (m^{-1}).
$n_2(x)$	particle size distribution of the catalyst in the product stream from the reactor, based on the number of particles: a function of x such that $n_2(x)\mathrm{d}x$ is equal to the fraction of the number of catalyst particles in the product stream from the reactor with a particle size between x and $(x + \mathrm{d}x)$, (m^{-1}).
$\mathbb{P}$	catalyst reactor productivity: total quantity of polymer produced by one *kg* of catalyst that has left the reactor, (*kg* polymer/*kg* catalyst).
$\mathcal{P}$	catalyst particle productivity: quantity of polymer produced by one catalyst particle during its lifetime at a certain concentration profile, (*kg* polymer/*kg* catalyst).
PV	pore volume of a silica gel catalyst support particle, or silica gel supported catalyst particle, (cm^3, g^{-1}).
Rx	symbol representing all of the reactor conditions, including temperature, concentrations, and flows.
SA	surface area of a silica gel catalyst support particle, or silica gel supported catalyst particle, (m^2, g^{-1}).
t	catalyst lifetime: the time interval between the moment when a catalyst particle is fed to the reactor and any moment at which the reactor system is considered, (*s*).
$W(x)$	cumulative particle size distribution of a catalyst or polymer powder: a function of x such that $W(x)$ is equal to the weight fraction of the powder with a particle size smaller than x, (–).
$w_0(x)$	particle size distribution of the catalyst in the catalyst feed to the reactor: a function of x such that $w_0(x)\mathrm{d}x$ is equal to the weight fraction of the catalyst feed with a particle size between x and $(x + \mathrm{d}x)$, (m^{-1}).

$w_2(x)$	particle size distribution of the catalyst leaving the reactor: a function of x such that $w_2(x)dx$ is equal to the weight fraction of that catalyst with a particle size between x and $(x + dx)$, (m^{-1}).
x	particle size of a catalyst or polymer powder, (m).
α_0	total number of catalyst particles comprised in 1 *kg* of catalyst feed, (–).
α_2	total number of catalyst particles comprised in 1 *kg* of catalyst in the product stream from the reactor, (–).
α_{12}	number of particles comprised in the fraction of a catalyst or polymer powder with particle size x between x_1 and x_2, (–).
$\alpha_{2,\mathrm{pol}}$	total number of polymer particles comprised in 1 *kg* of polymer in the product stream from the reactor, (–).
θ	catalyst residence time: total time that a catalyst particle has spent in the reactor at the moment when it leaves the reactor, (s).
ρ	particle density of a catalyst or polymer powder, (kg, m^{-3}).
ρ_{cat}	density of the porous silica or catalyst particle, (kg, m^{-3}).
$\rho_{\mathrm{cat,\,BD}}$	bulk density of the silica or catalyst powder, (kg, m^{-3}).

7 Catalyst Residence Time Distribution

In this chapter the catalyst residence time distribution (RTD) is calculated, of which the knowledge is vital to understand the functioning of a polyolefin polymerisation reactor. Indeed, the production rate can only be investigated, understood, or predicted by combining a catalyst residence time distribution with a catalyst activity profile. And inversely, data on catalyst activity can only be derived from full scale operation when the catalyst residence time distribution is known.

The discussion is started from statements that are as general as possible, in order to cope with the fact that a catalyst does not consist of a number of identical particles, but owns particles of different sizes, pore size distributions, and chemical loads. Still, it will result in a number of simple and practical statements due to the use of average values, and to the important assumption of perfect mixing for the catalyst.

In the discussion, abstraction is made of the liquid phase in the reactor. When the contents of the reactor are mentioned, the solid phase, i.e. catalyst particles with accumulated polymer, is meant. The reactor is seen as a mass of solids, with a feed stream of solid catalyst and a product stream of catalyst and polymer. The liquid is of course necessary as a medium to carry this solid phase, but does not interfere directly in the discussion. Where the liquid plays a role, by concentration of reagents and solids, all of its influence is wrapped up in the general term *reactor conditions*, and is considered to be a constant.

A number of definitions and concepts have already been introduced in the previous chapters. They are repeated here, and with some more detail, like e.g. their dependency on the catalyst particle size, and perhaps other catalyst properties.

7.1 Definitions and Assumptions

Most catalysts are produced by impregnation of silica carrier particles with active chemicals. Those carriers own a certain particle size distribution. Many properties of the catalyst, like responses to concentrations of the reactants, or catalyst activity and productivity, will depend on the catalyst particle size. This may be due to different diffusion paths to be followed during the polymerisation, but also simply to inhomogeneous impregnation of the silica particles.

Therefore, in a first and most general approach, catalyst activity and catalyst residence time will be considered as dependent on the *catalyst particle size*, x.

The *particle size distribution* of the catalyst *in the catalyst feed to the reactor*, $w_0(x)$, is a function of x such that $w_0(x)\mathrm{d}x$ is equal to the weight fraction of the catalyst feed with a particle size between x and $(x + \mathrm{d}x)$[1].

1 The subscript 0 refers to properties of the feeds to the reactor; the subscript 1 to the contents of the reactor; the subscript 2 to streams leaving the reactor.

The *particle size distribution* of the catalyst *present in the reactor*, $w_1(x)$, is a function of x such that $w_1(x)\mathrm{d}x$ is equal to the weight fraction of that catalyst with a particle size between x and $(x + \mathrm{d}x)$.
Likewise, $w_2(x)$ is the *particle size distribution* of the catalyst *leaving* the reactor.

Differences in particle size probably explain most of the variations in properties within one catalyst batch. However, if needed, to keep the discussion as general as possible, the expressions obtained in this chapter could be extended to more catalyst properties by simple analogy, adding, besides particle size distributions, more distribution functions for more properties.

If this approach offers the advantage of a basic and thorough understanding, it also leads to complex formulations, that are not very practical. However, it will be shown that the results are greatly simplified, and become really useful by the assumption of *perfect mixing of the catalyst in the loop reactor*. This assumption is clarified and justified in the next section.

The *catalyst residence time*, θ, is a property of a catalyst particle that has left the reactor, and is defined as the time interval between the moment when the catalyst particle has entered the reactor, and the moment when that catalyst particle has left the reactor.
The catalyst residence time distribution $E_2(\theta)$ is a function of the catalyst residence time θ such that $E_2(\theta)\mathrm{d}\theta$ is equal to the weight fraction of the catalyst with a residence time between θ and $(\theta + \mathrm{d}\theta)$.
The catalyst residence time distribution $E_2(\theta)$ is assumed to depend on the *reactor conditions*, Rx, and the catalyst particle size, x.
The symbol Rx is representing all of the reactor conditions, including reactor configuration and geometry, temperature, concentrations, masses in the reactor and flows.

The *catalyst lifetime*, t, is a property of a catalyst particle that is present in the reactor and is equal to the time interval between the moment when the catalyst particle has entered the reactor, and any moment at which the reactor system is considered.
The *catalyst lifetime distribution*, $E_1(t)$, is a function of the catalyst lifetime t such that $E_1(t)\mathrm{d}t$ is equal to the weight fraction of the catalyst with a lifetime between t and $(t + \mathrm{d}t)$.
The catalyst lifetime t is assumed to depend on the reactor conditions Rx, and the catalyst particle size x.

Since the polymer is constantly accumulated at the particle by polymerisation, one cannot assign a specific residence time nor lifetime to a *polymer* particle. Therefore the concept of *polymer residence time* is not meaningful and thus not used here.

Only the steady state is considered. By *steady state* is meant that the reactor as a whole is operated under stable conditions, and in particular that mass streams (production rate, feeds ...) and mass contents (masses in the reactor, in settling legs of a loop reactor...) are constant in time, but also that particle size distributions in those masses and streams are constant in time.

Cyclic phenomena, on a time scale that is small compared to the average catalyst residence time, and on a scale of space that is small compared to the reactor, like the functioning of the settling legs, or the mixing in the reactor on a micro-scale, are supposed to happen in constant cycles, and not in opposition to the steady state assumption.

The 1/1 hypothesis is assumed to be valid.

7.2 Perfect Mixing of the Catalyst

An important assumption is now made, which will greatly simplify the calculations in the next sections.

Stirred tank reactors and fluidized bed, or so called gas phase, reactors, are rightly considered to be perfectly mixed. However, for slurry loop reactors, equipped with settling legs to take off the polymer, it is generally believed that segregation of polymer particles is occurring during the sedimentation of the polymer particles in the settling legs, and consequently, that there is no perfect mixing of the polymer particles. It is now stated that even if there is segregation for polymer particles, this is not the case for catalyst particles as such; that *perfect mixing applies to the catalyst particles in the reactor, even if it does not apply to the polymer particles formed around these catalyst particles.*

Perfect mixing means that the contents of the reactor are perfectly mixed, that no segregation occurs where the product stream is taken from the reactor, and that the composition of the product stream is identical to the composition of the contents of the reactor.

This is assumed for catalyst particles, which are considered irrespective of the polymer. Abstraction should be made of the polymer around the catalyst particle, and only the catalyst particle should be considered. Of all catalyst fed to the reactor, no catalyst particles will be taken from the reactor preferentially, due to differences in catalyst particle size, lifetime nor any other catalyst property.

This assumption is clarified as follows.

1. A large fraction of the solids removed from a loop reactor, equipped with settling legs, will enter the settling legs by the mechanism of "displacement". For this fraction perfect mixing is self-evident.
   For the best settling powders, this fraction is, as calculated from the settling formulas, between 50 and 60%. For poorly settling powders it is as high as 75%.
2. The second fraction of the solids, from 25 to 50%, will enter the legs by the mechanism of "sedimentation", where segregation can prevail based on the size of the polymer particle. However, this "settling" is not to be understood as undisturbed settling through an immobile medium. "Settling" will also occur by particle collisions and turbulence. Bigger particles will even own a greater inertia when passing the settling leg nozzles, and in that respect be less prone to settle. Also, the particle morphology will be of influence.

3. For both loop and gas phase reactors, there is no direct correlation between catalyst particle size and polymer particle size.
   The polymer particle size is of an order of magnitude 10 times larger than the catalyst particle size, and is rather created by catalyst lifetime and activity. Small catalyst particles will often be more active and grow faster.
4. Catalyst lifetime, which, to support the assumption, should neither lead to segregation, is not too narrowly correlated with polymer particle size. The powder produced in a bench test will illustrate this, as, even if it is produced at a uniform residence time, it still owns a particle size distribution of a certain broadness.
5. The statement has been confirmed by experimental evidence (see below, Section 7.8).

A number of important consequences follow from the assumption of perfect mixing of the catalyst. Due to perfect mixing of the catalyst there will be no difference between the contents of the reactor and the product stream with respect to the catalyst, even if there may be differences in polymer particle size distribution. The obvious difference in solids concentration between the settled slurry in the legs and the bulk of the reactor does not play a role as the streams of solids are considered, regardless of the liquid medium.

The particle size distributions of the catalyst in the feed, in the reactor and in the product stream will be identical,

$$w_0(x) = w_1(x) = w_2(x) \,.$$

Catalyst lifetime and catalyst residence time will be identical,

$$t = \theta \,,$$

and so will be lifetime and residence time distributions,

$$E_1(t, Rx, x) = E_2(\theta, Rx, x) \,.$$

On top of that, catalyst lifetime and residence time distributions are independent of any catalyst property, as the assumptions imply that every weight fraction of catalyst leaving the reactor will contain a representative sample of all catalyst particle sizes, or other properties,

$$E_1(t, Rx, x) = E_2(\theta, Rx, x) = E_1(t, Rx) = E_2(\theta, Rx) \,.$$

7.3 Catalyst Activity

The catalyst activity is a property of the catalyst during its lifetime inside the reactor.

The *catalyst activity*, $\mathcal{A}(t, Rx, x)$, (*kg* polymer per *kg* of catalyst and per unit time), is the quantity of polymer produced by one catalyst particle or by a number of identical catalyst particles per unit time at some moment during its lifetime in the reactor. This activity will, in general, decline as lifetime proceeds, probably because the active sites get hemmed in by the growing polymer, and diffusion to the sites becomes more and more hindered. The catalyst activity is assumed to depend on the reactor conditions Rx, the catalyst particle size x, and the lifetime t of the catalyst in the reactor.

In reactors with a uniform concentration of ethylene as the only reactant, the catalyst activity $\mathcal{A}(t, Rx, x)$ is equal to the reaction rate r_2, equation (5.1), $r_2 = k_2(T)\mathrm{e}^{-at}y_2$. In a real life loop reactor the activity will often be the sum of three reaction rates, as, besides ethylene, also a comonomer (1-hexene, or another α-olefin) and hydrogen may participate in the polymerisation.

Also in the loop reactor the concentrations of the reactants are not uniform throughout the loop volume, as there is usually only one feeding point where the reactants are fed, whilst they are consumed by the reaction all along the length of the loop, before more reactant is added when the feeding point has been reached again. So the reactor conditions Rx stand for a concentration profile of up to three reactants along the loop, and not for a uniform concentration.

When the catalyst is used in a particular reactor, the *catalyst reactor activity*, $\mathbb{A}$, is defined as the rate of polymer produced per *kg* of all catalyst present in that reactor. This activity is the combined result of the activities of different fractions of catalyst present in the reactor at the same time, though at different lifetimes t, and will depend on the catalyst lifetime distribution.

$E_1(t)$ is the distribution function of the catalyst lifetime t. The fraction of catalyst present in the reactor, that owns a particle size x and a lifetime t is equal to,

$$w_1(x)\mathrm{d}xE_1(t, Rx, x)\mathrm{d}t\ .$$

The quantity of polymer produced per unit of time by this fraction of the catalyst in the reactor, with particle size x and lifetime t is,

$$\mathrm{d}\mathbb{A} = w_1(x)\mathrm{d}xE_1(t, Rx, x)\mathrm{d}t\mathcal{A}(t, Rx, x)\ .$$

The catalyst reactor activity is obtained by summing over all values of particle size and lifetime,

$$\mathbb{A} = \int_{x=0}^{\infty} w_1(x)\left[\int_{t=0}^{\infty} E_1(t, Rx, x)\mathcal{A}(t, Rx, x)\mathrm{d}t\right]\mathrm{d}x\ . \tag{7.1}$$

More catalyst properties, besides particle size, could have been taken into account. The distributions of these properties should have been defined then, and the activities $\mathrm{d}\mathbb{A}$ should have been summed over these properties too. As many more integral signs should have appeared here as more properties would have been added. The reasoning followed, however, would remain unchanged.

The assumption of perfect mixing for the catalyst is now applied, from which it follows that $t = \theta$, $w_1(x) = w_0(x)$ and $E_1(t, Rx, x) = E_2(\theta, Rx)$.

The reactor activity will be,

$$\mathbb{A} = \int_{\theta=0}^{\infty} E_2(\theta, Rx) \left[\int_{x=0}^{\infty} w_0(x) \mathcal{A}(\theta, Rx, x) \mathrm{d}x \right] \mathrm{d}\theta \,.$$

The integral at the right side is the *average activity of the catalyst over all catalyst particle sizes*, (or any other catalyst property),

$$\overline{\mathcal{A}}(\theta, Rx) = \int_{x=0}^{\infty} w_0(x) \mathcal{A}(\theta, Rx, x) \mathrm{d}x \,.$$

This concept $\overline{\mathcal{A}}$ absorbs all distributions of particle size or other properties.
In reactors with a uniform concentration of ethylene as the only reactant, $\overline{\mathcal{A}}(\theta, Rx)$ would be equal to the *the catalyst activity profile*, as recorded in micro scale reactor tests with uniform catalyst residence times. The activity profile in the loop reactor is somewhat different from what is measured in the lab because more reactants are involved than just ethylene, and the concentration profile along the loop is not uniform. So the average catalyst activity is not only an average over the catalyst properties, but also over the concentration profiles met in the loop reactor. This full scale activity profile can be measured, as explained in the next Chapter 8.
The catalyst reactor activity is thus,

$$\mathbb{A} = \int_{\theta=0}^{\infty} E_2(\theta, Rx) \overline{\mathcal{A}}(\theta, Rx) \mathrm{d}\theta \,. \tag{7.2}$$

7.4 Catalyst Reactor Productivity

Contrary to the catalyst activity, the catalyst reactor productivity is a property of the catalyst that has left the reactor.

The *catalyst reactor productivity*, $\mathbb{P}$, (*kg* polymer per *kg* catalyst) is the total quantity of polymer produced by one *kg* of catalyst that has left the reactor.

The fraction of the catalyst feed, that owns a particle size x and a residence time θ, is equal to,

$$w_0(x) \mathrm{d}x E_2(\theta, Rx, x) \mathrm{d}\theta \,.$$

The quantity of polymer produced by this fraction of the catalyst feed, with particle size x and residence time θ is the sum of all polymer produced by that particular catalyst fraction during the time it spent in the reactor, with a lifetime going from 0 to θ,

$$\mathrm{d}\mathbb{P} = w_0(x) \mathrm{d}x E_2(\theta, Rx, x) \mathrm{d}\theta \int_{t=0}^{\theta} \mathcal{A}(t, Rx, x) \mathrm{d}t \,.$$

The catalyst reactor productivity is obtained by summing over all values of particle size and residence time,

$$\mathbb{P} = \int_{x=0}^{\infty} w_0(x) \left\{ \int_{\theta=0}^{\infty} E_2(\theta, Rx, x) \left[\int_{t=0}^{\theta} \mathcal{A}(t, Rx, x)\mathrm{d}t \right] \mathrm{d}\theta \right\} \mathrm{d}x\,. \tag{7.3}$$

Applying the assumption of perfect mixing for the catalyst, $E_2(\theta, Rx, x) = E_2(\theta, Rx)$. The catalyst reactor productivity will be,

$$\mathbb{P} = \int_{\theta=0}^{\infty} E_2(\theta, Rx) \left\{ \int_{x=0}^{\infty} w_0(x) \left[\int_{t=0}^{\theta} \mathcal{A}(t, Rx, x)\mathrm{d}t \right] \mathrm{d}x \right\} \mathrm{d}\theta\,,$$

or,

$$\mathbb{P} = \int_{\theta=0}^{\infty} E_2(\theta, Rx) \left[\int_{t=0}^{\theta} \overline{\mathcal{A}}(t, Rx)\mathrm{d}t \right] \mathrm{d}\theta\,. \tag{7.4}$$

The integral on the right side is nothing else than the catalyst particle productivity, as defined by equation (6.1),

$$\mathcal{P}(\theta, Rx) = \int_{t=0}^{\theta} \overline{\mathcal{A}}(t, Rx)\mathrm{d}t\,.$$

It is not surprising that the catalyst reactor productivity can be calculated as the sum of all catalyst particle productivities for all fractions of different residence times, since the total mass of polymer is the sum of all polymer produced on the individual catalyst particles, or,

$$\mathbb{P} = \int_{\theta=0}^{\infty} E_2(\theta, Rx)\mathcal{P}(\theta, Rx)\mathrm{d}\theta\,. \tag{7.5}$$

7.5 Mass of Polymer Present in the Reactor

Let m_{cat} represent the *total mass of catalyst present in the reactor.*
According to the 1/1 hypothesis, the *total mass of polymer in the reactor*, m_{pol}, is equal to the total mass of polymer produced by all catalyst particles present in the reactor. The quantity of polymer, present in the reactor and produced by the fraction of the catalyst feed, which owns a particle size x and lifetime t, is,

$$\mathrm{d}m_{\mathrm{pol}} = m_{\mathrm{cat}} w_1(x)\mathrm{d}x E_1(t, Rx, x)\mathrm{d}t \int_{t'=0}^{t} \mathcal{A}(t', Rx, x)\mathrm{d}t'\,.$$

The total mass of polymer in the reactor is obtained by summing over all values of particle size and residence time,

$$m_{\text{pol}} = m_{\text{cat}} \int_{x=0}^{\infty} w_1(x) \left\{ \int_{t=0}^{\infty} E_1(t, Rx, x) \left[\int_{t'=0}^{t} \mathcal{A}(t', Rx, x)\mathrm{d}t' \right] \mathrm{d}t \right\} \mathrm{d}x\,. \tag{7.6}$$

It is worth noting that this expression is formally similar to expression (7.3) for catalyst productivity, but of basically different contents.

Again, under perfect mixing for the catalyst, $t = \theta$, $w_1(x) = w_0(x)$ and $E_1(t, Rx, x) = E_2(\theta, Rx)$, and the mass of polymer in the reactor will be,

$$m_{\text{pol}} = m_{\text{cat}} \int_{\theta=0}^{\infty} E_2(\theta, Rx) \left\{ \int_{x=0}^{\infty} w_0(x) \left[\int_{t=0}^{\theta} \mathcal{A}(t, Rx, x)\mathrm{d}t \right] \mathrm{d}x \right\} \mathrm{d}\theta\,,$$

or,

$$m_{\text{pol}} = m_{\text{cat}} \int_{\theta=0}^{\infty} E_2(\theta, Rx) \left[\int_{t=0}^{\theta} \overline{\mathcal{A}}(t, Rx)\mathrm{d}t \right] \mathrm{d}\theta\,.$$

The similarity to expression (7.3) for catalyst productivity, has now become of a more than formal nature and, by comparison to (7.4),

$$m_{\text{pol}} = m_{\text{cat}}\mathbb{P}\,. \tag{7.7}$$

It is thus concluded that *under steady state conditions, the 1/1 hypothesis and perfect mixing of the catalyst, the mass of polymer present in the reactor is equal to the mass of catalyst in the reactor times the catalyst productivity.*

7.6 Catalyst Residence Time Distribution

The calculation of the residence time distribution for a perfectly mixed reactor, applied here to the catalyst in the loop reactor, is well known from chemical reactor theory.

It is assumed that a small fraction of the *catalyst feed flow*, F_{cat}, is marked by some tracer property, injected at time $\theta = 0$ in the reactor, and followed on its path through the reactor. Let its *quantity* be $q_0 = F_{\text{cat}}\mathrm{d}\theta$.

Let the *quantity of the tracer catalyst present in the reactor* be q_1.

The concentration of the tracer catalyst in the polymer in the reactor is, making use of (7.7), equal to,

$$\frac{q_1}{m_{\text{pol}}} = \frac{q_1}{m_{\text{cat}}\mathbb{P}}\,.$$

The stream of polymer leaving the reactor under steady state conditions is $F_{cat}\mathbb{P}$. A mass balance for the tracer catalyst, taking into account that all catalyst is present in the solid (polymer) phase, is,

$$-dq_1 = \frac{q_1}{m_{cat}\mathbb{P}} F_{cat}\mathbb{P}d\theta \,,$$

or,

$$\frac{dq_1}{q_1} = -\frac{F_{cat}}{m_{cat}} d\theta \,,$$

or, after integration on both sides,

$$\log q_1 = -\frac{F_{cat}}{m_{cat}}\theta + c \,,$$

which is solved, with $q_1(\theta = 0) = q_0$, as,

$$q_1 = q_0 \exp\left(\frac{-F_{cat}}{m_{cat}}\theta\right) .$$

The catalyst residence time distribution $E_2(\theta)$ is a function of θ such that $E_2(\theta)d\theta$ is equal to the weight fraction of the catalyst with a residence time between θ and $(\theta + d\theta)$,

$$E_2(\theta)d\theta = \frac{-dq_1(\theta)}{q_0} = \frac{1}{q_0} q_1 \frac{F_{cat}}{m_{cat}} d\theta \,.$$

Hence,

$$E_2(\theta) = \frac{F_{cat}}{m_{cat}} \exp\left(\frac{-F_{cat}}{m_{cat}}\theta\right) . \tag{7.8}$$

The *average catalyst residence time* τ is calculated as,

$$\tau = \int_{\theta=0}^{\infty} \theta E_2(\theta)d\theta = \frac{m_{cat}}{F_{cat}} \,,$$

using the method of integration by parts, with $u = \theta$ and $dv = E_2(\theta)d\theta$[2]. The average catalyst residence time τ can also be expressed in terms of the polymer as,

$$\tau = \frac{m_{cat}}{F_{cat}} = \frac{m_{cat}\mathbb{P}}{F_{cat}\mathbb{P}} = \frac{m_{pol}}{P} \,,$$

2 Let $u = \theta$, then $du = d\theta$. Let $dv = E_2(\theta)d\theta = F_{cat}/m_{cat} \exp((-F_{cat}/m_{cat})\theta)d\theta$, then $v = -\exp((-F_{cat}/m_{cat})\theta)$.

$$\int_{\theta=0}^{\infty} \theta E_2(\theta)d\theta = \int_{\theta=0}^{\infty} u dv = [uv]_{\theta=0}^{\infty} - \int_{\theta=0}^{\infty} v du = \left[-\theta \exp\left(\frac{-F_{cat}}{m_{cat}}\theta\right)\right]_{\theta=0}^{\infty} + \int_{\theta=0}^{\infty} \exp\left(\frac{-F_{cat}}{m_{cat}}\theta\right) d\theta \,,$$

which is calculated as m_{cat}/F_{cat}. A problem is raised by the term $[-\theta \exp((-F_{cat}/m_{cat})\theta)]_{\theta=\infty}$, which yields the undefined form $0 \cdot \infty$. It can be shown to be zero by application of L'Hospital's rule.

where P is the *polymer production rate*. This expression is formally equal to an *average polymer residence time*, and so it is often called, though this term is physically rather meaningless, since a polymer particle is growing continuously, and no definite residence time can be assigned to the polymer.

The *catalyst residence time distribution* has thus been found as,

$$E_2(\theta) = \frac{1}{\tau}\exp\left(\frac{-\theta}{\tau}\right),$$
$$\tau = \frac{m_{\text{cat}}}{F_{\text{cat}}} = \frac{m_{\text{pol}}}{P}\,. \tag{7.9}$$

The average catalyst residence time, and thus the residence time distribution, is a very accessible quantity, as both m_{pol} and P are easily available from the reactor process computer[3].

7.7 Mass Balance for the Polymer

Under steady state conditions, the polymer rate P produced in the reactor is equal to the stream of polymer leaving the reactor,

$$P = m_{\text{cat}}\mathbb{A} = F_{\text{cat}}\mathbb{P}\,. \tag{7.10}$$

When perfect mixing for the catalyst holds, equation (7.10) together with (7.9) yields the simple relation between production rate, catalyst productivity and catalyst reactor activity,

$$\frac{P}{F_{\text{cat}}} = \frac{m_{\text{cat}}}{F_{\text{cat}}}\mathbb{A} = \mathbb{P}\,,$$
$$\mathbb{P} = \frac{P}{F_{\text{cat}}} = \tau\mathbb{A}\,. \tag{7.11}$$

By introducing the expression (7.9) for the catalyst residence time distribution, together with the equations (7.2) and (7.4), in the polymer mass balance (7.10), two equivalent expressions for the catalyst productivity are obtained,

$$\int_{\theta=0}^{\infty}\exp\left(\frac{-\theta}{\tau}\right)\overline{\mathcal{A}}(\theta, Rx)\mathrm{d}\theta \equiv \int_{\theta=0}^{\infty}\frac{1}{\tau}\exp\left(\frac{-\theta}{\tau}\right)\left[\int_{t=0}^{\theta}\overline{\mathcal{A}}(t, Rx)\mathrm{d}t\right]\mathrm{d}\theta = \mathbb{P}\,.$$

It can be verified mathematically that this is an identity. Both members of the identity show how a certain catalyst productivity will be reached in the steady state by mutual

3 The production rate P is calculated from a mass balance over the reactor, or from a precise energy balance. The mass of polymer is calculated from the reactor volume, the solids concentration and the polymer density, $m_{\text{pol}} = V_{\text{R}}C_0\rho_{\text{p}}$.

interaction of the catalyst activity profile and the catalyst residence time distribution. The solids concentration in the reactor e.g., will increase if the catalyst activity profile is higher, but a higher solids concentration will also increase the average catalyst residence time.

The right member of the identity is the expression (7.5) for the catalyst reactor productivity in terms of the particle productivity,

$$\mathbb{P} = \int_{\theta=0}^{\infty} E_2(\theta, Rx) \left[\int_{t=0}^{\theta} \overline{\mathcal{A}}(t, Rx)\mathrm{d}t \right] \mathrm{d}\theta = \int_{\theta=0}^{\infty} E_2(\theta, Rx)\mathcal{P}(\theta, Rx)\mathrm{d}\theta \,.$$

The left member of the identity,

$$\mathbb{P} = \int_{\theta=0}^{\infty} \exp\left(\frac{-\theta}{\tau}\right) \overline{\mathcal{A}}(\theta, Rx)\mathrm{d}\theta \,, \tag{7.12}$$

will be found to be extremely useful to calculate catalyst activity profiles from commercial loop reactor data in the next Chapter 8.

Note: Production Rate at Constant Catalyst Activity
In the hypothetical case of a constant catalyst activity, the catalyst reactor activity $\mathbb{A}$ is equal to the catalyst activity $\overline{\mathcal{A}}$, and the production rate can be expressed as follows.
From equation (7.11),

$$P = F_{\mathrm{cat}}\mathbb{P} = F_{\mathrm{cat}}\overline{\mathcal{A}}\tau \,.$$

From equation (7.9),

$$\tau = \frac{m_{\mathrm{pol}}}{P} = \frac{V_{\mathrm{R}} C_0 \rho_{\mathrm{p}}}{P} \,.$$

It follows that,

$$P = (F_{\mathrm{cat}}\overline{\mathcal{A}} V_{\mathrm{R}} C_0 \rho_{\mathrm{p}})^{0.5} \,. \tag{7.13}$$

This hypothetical example shows that the production rate is not growing linearly with the catalyst feed rate, because a higher production rate is reducing the catalyst residence time. It is not always well understood that for running high production rates in the loop reactor, a toll must be paid in higher catalyst consumptions. □

7.8 Experimental Measurement of the Catalyst Residence Time Distribution in the Loop Reactor

The catalyst residence time distribution has been measured by the author in a full-scale commercial reactor, by making use of the reactor's response to a step signal in the catalyst feed. The assumptions and calculations made in this chapter have been fully confirmed.

7.8.1 Cumulative RTD and Response to a Step Signal

The use of a step signal to determine a residence time distribution in a reactor from experiment, is also well known in chemical reactor theory.
The *cumulative catalyst residence time distribution*, $F_2(\theta)$, is a function of the catalyst residence time θ such that $F_2(\theta)$ is equal to the weight fraction of the catalyst with a residence time between 0 and θ.
By definition,

$$F_2(\theta) = \int_{\theta'=0}^{\theta} E_2(\theta')\mathrm{d}\theta' .$$

If the residence time distribution, equation (7.9), holds, then,

$$F_2(\theta) = \int_{\theta'=0}^{\theta} \frac{1}{\tau}\exp\left(\frac{-\theta'}{\tau}\right)\mathrm{d}\theta' = 1 - \exp\left(\frac{-\theta}{\tau}\right) . \tag{7.14}$$

Suppose a step signal is applied at the time $\theta = 0$, from where all fed catalyst is marked by a certain concentration of a tracer substance.
The weight fraction of the catalyst leaving the reactor with a residence time longer than θ, i.e. of the unmarked catalyst, is equal to $[1 - F_2(\theta)]$.
Suppose s_0 is the weight fraction of the tracer in the catalyst fed before the step change;
s_1 is the weight fraction of the tracer in the catalyst fed after the step change;
and s is the weight fraction of the tracer in the catalyst leaving the reactor.
The catalyst leaving the reactor is a blend of unmarked and marked catalyst, so,

$$s = [1 - F_2(\theta)]s_0 + F_2(\theta)s_1 ,$$

or,

$$F_2(\theta) = \frac{s - s_0}{s_1 - s_0} . \tag{7.15}$$

By measuring the tracer content in the catalyst leaving the reactor, (or product, if the catalyst productivity does not change), $F_2(\theta)$ is obtained and thus $E_2(\theta)$.

7.8.2 Experiment

The catalyst residence time distribution has been measured by the author in a full-scale commercial loop reactor. Advantage was taken of an occasion where a chromium oxide catalyst without Ti was partially replaced by another chromium oxide catalyst containing Ti. The replacement was executed at once, so it could be interpreted as a step change experiment, where Ti served as the tracer element. The reactor conditions, especially solids concentration and catalyst productivity, were kept as constant as possible before and after the step change.

This test had been prepared by an earlier similar test, which did not succeed, but offered the necessary experience to make the second trial successful. Critical are the correct timing of the product sampling after the step signal has been applied, and, above all, the skill of the lab technician to analyse Ti and Cr in the polymer samples at a level of 1 *ppm*.

The response curve *s*, and the average catalyst residence time according to (7.9), $\tau = m_{pol}/P$, (here labeled as polymer residence time), were obtained from the experiment, both as a function of time, as shown in Figure 7.1. Time was expressed in minutes.

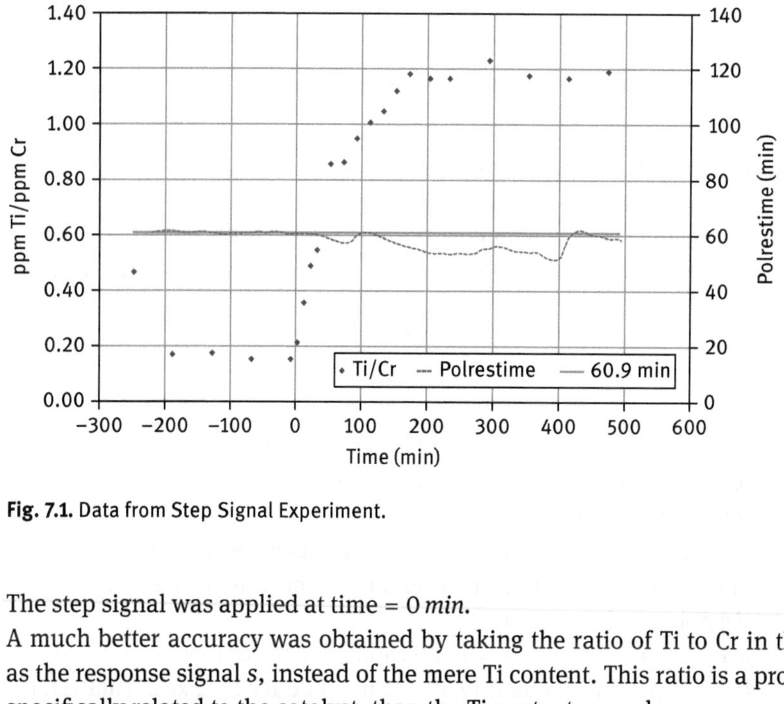

Fig. 7.1. Data from Step Signal Experiment.

The step signal was applied at time = 0 *min*.

A much better accuracy was obtained by taking the ratio of Ti to Cr in the polymer as the response signal *s*, instead of the mere Ti content. This ratio is a property more specifically related to the catalyst, than the Ti content as such.

The cumulative residence time distribution function $F_2(\theta)$ was calculated from the data points according to (7.15). Assuming that an equation of the form of equation (7.14) was valid for $F_2(\theta)$, and using a least squares method, a value of the average catalyst residence time τ was fitted to the data, as shown in Figure 7.2. The data points fitted the presumed curve very well. The experimental value found for τ was exactly equal to the value of $\tau = m_{pol}/P = 60.9$ at the moment of the step change (time = 0), and in the early minutes afterwards, i.e. in the time span which is most decisive to the shape of the curve, as seen in Figure 7.1. The response curve did thus match the equation (7.9) exactly.

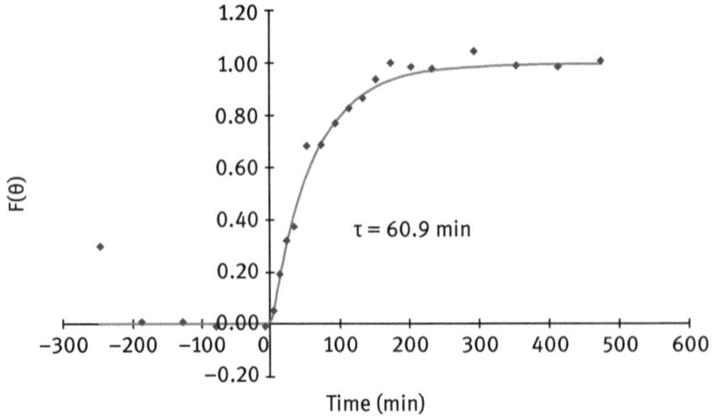

Fig. 7.2. Cumulative RTD function derived from Step Signal Experiment.

The experiment confirms the calculations presented in this chapter, and the assumption of perfect mixing for the catalyst. It is another example of how interesting information can be obtained from a commercial scale reactor, if use is made of the opportunities offered by the production programme and, especially, of large scale testruns.

Notations

$\mathbb{A}$	catalyst reactor activity: quantity of polymer produced by all catalyst present in the reactor per unit time, per *kg* catalyst, (*kg* polymer/*kg* catalyst, *s*).
$\mathcal{A}(t, Rx, x)$	catalyst activity: quantity of polymer produced by the catalyst, per *kg* of catalyst, per unit time, at a certain lifetime *t* in the reactor, (*kg* polymer/*kg* catalyst, *s*).
$\overline{\mathcal{A}}(t, Rx)$	average catalyst activity over all catalyst particle sizes (or any other catalyst property): quantity of polymer produced per *kg* of catalyst, per unit time, at a certain lifetime *t* in the reactor, (*kg* polymer/*kg* catalyst, *s*).
C_0	volumetric concentration of solids in the reactor, (–).
$E_1(t)$	catalyst lifetime distribution: a function of the catalyst lifetime *t* such that $E_1(t)\mathrm{d}t$ is equal to the weight fraction of the catalyst with a lifetime between *t* and $(t + \mathrm{d}t)$, (s^{-1}).
$E_2(\theta)$	catalyst residence time distribution: a function of the catalyst residence time θ such that $E_2(\theta)\mathrm{d}\theta$ is equal to the weight fraction of the catalyst with a residence time between θ and $(\theta + \mathrm{d}\theta)$, ($s^{-1}$).
F_{cat}	catalyst feed to the reactor, (kg, s^{-1}).

$F_2(\theta)$	cumulative catalyst residence time distribution: a function of the catalyst residence time θ such that $F_2(\theta)$ is equal to the weight fraction of the catalyst with a residence time between 0 and θ, (–).
m_{cat}	mass of catalyst present in the reactor, (*kg*).
m_{pol}	mass of polymer present in the reactor, (*kg*).
P	polymer production rate: total quantity of polymer produced in the reactor per unit of time, (*kg* polymer, s^{-1}).
$\mathbb{P}$	catalyst reactor productivity: total quantity of polymer produced by one *kg* of catalyst that has left the reactor, (*kg* polymer/*kg* catalyst).
q_0	a small quantity of tracer catalyst, injected at time $\theta = 0$ in the reactor, (*kg*).
q_1	quantity of the tracer catalyst q_0, present in the reactor, (*kg*).
Rx	symbol representing all of the reactor conditions, including temperature, concentrations, and flows.
s	weight fraction of tracer substance in the catalyst leaving the reactor during a step change tracer experiment, (–).
s_0	weight fraction of tracer substance in the catalyst fed before the step change in a step change tracer experiment, (–).
s_1	weight fraction of tracer substance in the catalyst fed after the step change in a step change tracer experiment, (–).
t	catalyst lifetime: the time interval between the moment when a catalyst particle that is present in the reactor, has entered the reactor, and any moment at which the reactor system is considered, (*s*).
V_R	reactor volume, (m^3).
$w_0(x)$	particle size distribution of the catalyst in the catalyst feed to the reactor: a function of x such that $w_0(x)dx$ is equal to the weight fraction of the catalyst feed with a particle size between x and $(x + dx)$, (m^{-1}).
$w_1(x)$	particle size distribution of the catalyst present in the reactor: a function of x such that $w_1(x)dx$ is equal to the weight fraction of that catalyst with a particle size between x and $(x + dx)$, (m^{-1}).
$w_2(x)$	particle size distribution of the catalyst leaving the reactor: a function of x such that $w_2(x)dx$ is equal to the weight fraction of that catalyst with a particle size between x and $(x + dx)$, (m^{-1}).
x	catalyst particle size, (*m*).
θ	catalyst residence time: the total time that a catalyst particle has spent in the reactor at the moment when it leaves the reactor, (*s*).
ρ_p	density of the polymer in the reactor, (*kg*, m^{-3}).
τ	average catalyst residence time, (*s*).

8 Catalyst Activity Profiles from Full Scale Loop Data

In this chapter is shown how data on catalyst activity can be derived from full scale loop reactor operation data.
The method is built on the results from the previous Chapter 7 on the catalyst residence time distribution, accepting the underlying definitions and assumptions.

8.1 Calculating the Catalyst Activity Profile from Full Scale Loop Data

Under the assumptions,
1. steady state,
2. the 1/1 hypothesis,
3. perfect mixing of the catalyst in the loop reactor,

a relation has been obtained in the previous Chapter 7, between the catalyst productivity and the catalyst activity profile in the loop reactor, equation (7.12),

$$\mathbb{P} = \int_{\theta=0}^{\infty} \exp\left(\frac{-\theta}{\tau}\right) \overline{\mathcal{A}}(\theta, Rx) d\theta \, .$$

The catalyst productivity $\mathbb{P}$ (*kg* polymer per *kg* catalyst) is easily obtained from full scale reactor data, where the production rate is routinely measured, if there is a sufficiently accurate measurement of the catalyst feed rate.
The catalyst activity profile in the reactor, $\overline{\mathcal{A}}(\theta, Rx)$, cannot be measured directly, but is, with the help of equation (7.12), derived from an empirical *productivity curve*, catalyst productivity versus residence time, $\mathbb{P}(\tau)$.
Indeed, remembering the definition of the *Laplace transform*[1] of a function $f(t)$,

$$\mathcal{L}\{f(t)\}(p) \equiv \int_{t=0}^{\infty} e^{-pt} f(t) \, ,$$

it is recognised from equation (7.12) that the catalyst productivity $\mathbb{P}(\tau)$ is the Laplace transform of the catalyst activity profile $\overline{\mathcal{A}}(\theta)$, with the variable p put equal to τ^{-1},

1 A function $f(t)$ is, by the Laplace transform, transformed into a function of the new variable p. The best known application of this transform is the solving of linear differential equations, which, by this transform, are turned into algebraic equations. The solution of the differential equation is then found as the inverse transform (if successful) of the solution of the algebraic equation. Here, the Laplace transform offers an elegant notation of the relation between the catalyst productivity curve and the activity profile, and an easy calculation of the latter from the former, since tables of (inverse) Laplace transforms are available in the literature.

and t equal to θ,

$$\mathbb{P}(\tau, Rx) = \mathcal{L}\{\overline{\mathcal{A}}(\theta, Rx)\}, p = \frac{1}{\tau} . \tag{8.1}$$

This result is not that surprising, as it is not unknown in the literature that the residence time distribution of a perfectly mixed stirred tank is functioning as a Laplace transformer of certain properties.

So, when, at a certain set of reactor conditions, a function $\mathbb{P}(\tau)$ has been measured experimentally, the catalyst activity profile $\overline{\mathcal{A}}(\theta)$ is obtained as the inverse Laplace transform of the catalyst productivity, with the variable p put equal to τ^{-1},

$$\overline{\mathcal{A}}(\theta) = \mathcal{L}^{-1}\left\{\mathbb{P}\left(p = \frac{1}{\tau}\right)\right\} . \tag{8.2}$$

This is illustrated by the following example.

Example

It is assumed in this example that the catalyst productivity at a certain set of reactor conditions is found to be a linear function of the catalyst residence time, $\mathbb{P} = b\tau$.

The catalyst activity profile is calculated as follows,

$$\mathbb{P} = b\tau = \frac{b}{\frac{1}{\tau}} = \frac{b}{p} .$$

With the help of a table of Laplace transforms is found,

$$\overline{\mathcal{A}}(\theta) = \mathcal{L}^{-1}\left\{\frac{b}{p}\right\} = b\mathcal{L}^{-1}\left\{\frac{1}{p}\right\} = b .$$

The catalyst activity profile is a constant.

Though this example is merely hypothetical, it already gives some guidance to draw conclusions from a first inspection of the catalyst productivity curve. If the catalyst productivity is increasing with the residence time at a rate less than linear, which is usually the case, then the catalyst activity profile is decreasing with catalyst lifetime. The opposite, an increasing activity profile, is sometimes, erroneously, concluded from results from certain bench scale reactor experiments. □

8.2 Catalyst Productivity Curve and Activity Profile in the Loop Reactor

A number of catalyst productivity curves, $\mathbb{P}(\tau)$, have been measured by the author, and the corresponding catalyst activity profiles derived therefrom, see Figures 8.1 to 8.5, where the units chosen for the residence time τ are (h), and for the catalyst productivity (kg polymer/ kg catalyst).

A least squares method was applied to fit an analytical expression to the experimental data points of the productivity curve. Since the catalyst cannot be productive at a zero residence time, the point (0, 0) was added each time to the set of experimental data points.

As a result it was found that all curves are fitted best by the expression,

$$\mathbb{P} = b\frac{\tau}{1 + a\tau}, \tag{8.3}$$

where the constants a and b are positive numbers. Their units are (s^{-1}), and (*kg* polymer/ *kg* catalyst, *s*) respectively.
So, on empirical grounds, it is stated that equation (8.3) *is, under the said assumptions, representing the productivity curve in an HDPE slurry loop reactor.*
The catalyst activity profile is derived from it, according to equation (8.2). In a table of Laplace transforms is found,

$$\mathcal{L}\{e^{at}\} = \frac{1}{p - a},$$

so the corresponding expression for the catalyst activity profile is,

$$\overline{\mathcal{A}}(\theta) = \mathcal{L}^{-1}\left\{b\frac{\tau}{1 + a\tau}\right\} = b\mathcal{L}^{-1}\left\{\frac{1}{\frac{1}{\tau} + a}\right\} = b\mathcal{L}^{-1}\left\{\frac{1}{p - (-a)}\right\} = be^{-a\theta}.$$

On the same empirical grounds and under the said assumptions, it is stated that the catalyst activity profile in an HDPE slurry loop reactor is,

$$\overline{\mathcal{A}}(\theta) = be^{-a\theta}. \tag{8.4}$$

This result corresponds very well with the results obtained in micro scale reactor tests with uniform catalyst residence times, as shown in Section 5.5. It is important to note that the catalyst activity is depending on the reactor conditions Rx, and that the measured activity profile is only valid for those reactor conditions.
The initial activity at $\tau = 0$ is b. The activity is declining as lifetime proceeds. The larger the value of the constant a, the more rapid is that decline.

The catalyst reactor activity can now also be quantified, from equations (7.11) and (8.3), as,

$$\mathbb{A} = \frac{\mathbb{P}}{\tau} = \frac{b}{1 + a\tau}. \tag{8.5}$$

A good understanding of the evolution of the productivity curve $\mathbb{P}(\tau)$ is useful. The productivity can be rewritten as,

$$\mathbb{P} = b\frac{\tau}{1 + a\tau} = b\tau - \frac{ab\tau^2}{1 + a\tau}.$$

At a given value of b, the maximum productivity is found when $a \to 0$, or when the catalyst shows no decline in activity with time. The productivity is then a linear relation equal to $\mathbb{P} = b\tau$.
In real life, catalysts are showing a decline of activity with time, expressed by the value of a. At higher values of a, the productivity curve will deviate more and more from the linear maximum towards lower values. Longer residence times do produce higher productivities, but the gain is less than linear.

The derivative of the productivity with respect to τ is,

$$\frac{\partial \mathbb{P}}{\partial \tau} = \frac{b}{1 + a\tau} - \frac{ab\tau}{(1 + a\tau)^2} = \frac{b}{(1 + a\tau)^2} \, .$$

At given values of a and b, the derivative of the productivity tends to zero when $\tau \to \infty$. Productivity will increase less and less with higher residence times, and in the end stay at a constant value, when the catalyst has lost all of its activity.
At a given value of b, the derivative of the productivity tends to zero when $a \to \infty$. This means that, at higher values of a, i.e. for catalysts with more rapidly declining activities, the gain in productivity with higher residence times will be smaller. Indeed, when a catalyst activity falls strongly with residence time, prolonged residence times add less and less to the total productivity. Such differences in productivity curves are seen between the different catalyst families, chromium oxide, Ziegler–Natta and metallocene.

8.3 Catalyst Particle Productivity

The catalyst particle productivity, $\mathcal{P}$, defined in Section 6.3, can now be quantified from Equation 6.1 as,

$$\mathcal{P} = \int_{t=0}^{\theta} \overline{\mathcal{A}}(t, Rx)\mathrm{d}t = \int_{t=0}^{\theta} b\mathrm{e}^{-at}\mathrm{d}t \, ,$$

$$\mathcal{P} = \frac{b}{a}\left(1 - \mathrm{e}^{-a\theta}\right) . \tag{8.6}$$

Equation 7.5, stating that the catalyst reactor productivity is equal to the sum of all particle productivities for all catalyst particles, is easily checked,

$$\mathbb{P} = \int_{\theta=0}^{\infty} E_2(\theta)\mathrm{d}\theta\mathcal{P}(\theta) = \int_{\theta=0}^{\infty} \frac{1}{\tau}\exp\left(-\frac{\theta}{\tau}\right)\frac{b}{a}\left(1 - \mathrm{e}^{-a\theta}\right)\mathrm{d}\theta = b\frac{\tau}{1 + a\tau} \, ,$$

which result is, as it should be, equal to expression (8.3).

The decline in catalyst activity with lifetime is explained by the polymerisation itself, because the active sites on the catalyst particle get hemmed in by the growing polymer, and diffusion to the sites becomes more and more hindered. This is illustrated by the following reasoning.
The particle productivity, $\mathcal{P}$, can be expressed in terms of the activity profile,

$$a\mathcal{P} = b - b\mathrm{e}^{-a\theta} = b - \mathcal{A} \, ,$$

and the activity profile in terms of the particle productivity,

$$\mathcal{A} = b - a\mathcal{P}\,.$$

Or, that the activity profile is exponentially declining with time, means that the activity of a catalyst particle is decreasing linearly with its own productivity, i.e. with the quantity of polymer amassed on its surface. This view is very close to the representation of active sites getting enclosed in the polymer.

8.4 Experimental Results for Catalyst Activity Profiles from Full Scale Data

A number of experiments, to measure a catalyst productivity curve $\mathbb{P}(\tau)$, and to deduct a catalyst activity profile therefrom, has been carried out by the author in a commercial slurry loop reactor.
To achieve a successful test, at least three different stable reactor conditions are required, at a constant reactor concentration profile, but at three different catalyst residence times. To realise the different catalyst residence times, the solids concentration in the loop is kept as constant as possible at a high concentration, whilst the production rate is varied. It is important to use stable conditions to obtain valuable data. The best way to achieve this, is to start from a high and stable production rate, and to decrease the catalyst feed in a discrete number of steps, allowing the reactor to stabilize after each step. The catalyst feed rate at each stage is estimated beforehand to yield, more or less, a certain production rate and catalyst residence time. Aiming directly at a precise production rate is not so practical and takes much more time before reaching stable conditions. Even at a constant catalyst feed rate, those will be reached only after some 12 hours. Each stage thus offers, after careful screening, a number of stable data points around a certain residence time. All other data points are omitted.
In order to obtain reliable data it is vital to have a precise measurement of the catalyst feed rate, and a proper method to calibrate it.

Figure 8.1 shows the results obtained for a Ziegler–Natta catalyst. The figure represents the experimental data points, the fitted productivity curve and the analytical activity profile $be^{-a\theta}$ derived therefrom. The constants are $a = 0.398$ and $b = 20926$. These values, as in all other graphs, have, for convenience, been expressed in (h^{-1}), and (kg polymer/ kg catalyst, h). Ziegler–Natta catalysts show a high productivity and a rather strong decline in activity.

Figure 8.2 shows the results obtained for a metallocene catalyst. The constants are $a = 1.000$ and $b = 7874$. Metallocene catalysts show a quite high initial activity, which is declining very rapidly with lifetime. This makes the operation of a metallocene catalyst in a loop reactor more risky, because strong local differences in ethylene

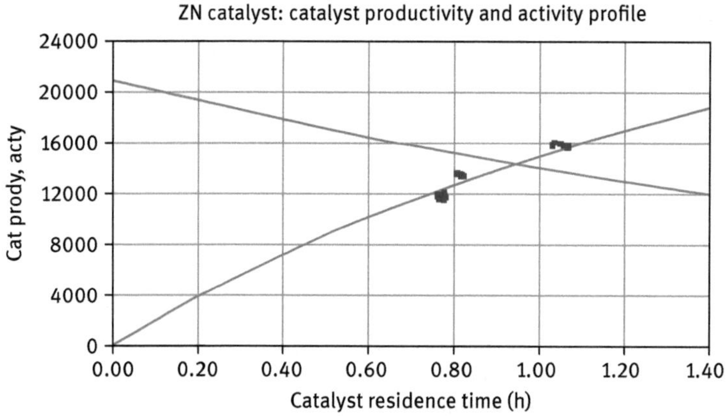

Fig. 8.1. Experimental Catalyst Productivity v. Residence Time curve and derived Catalyst Activity Profile for a Ziegler–Natta catalyst; a=0.398; b=20 926.

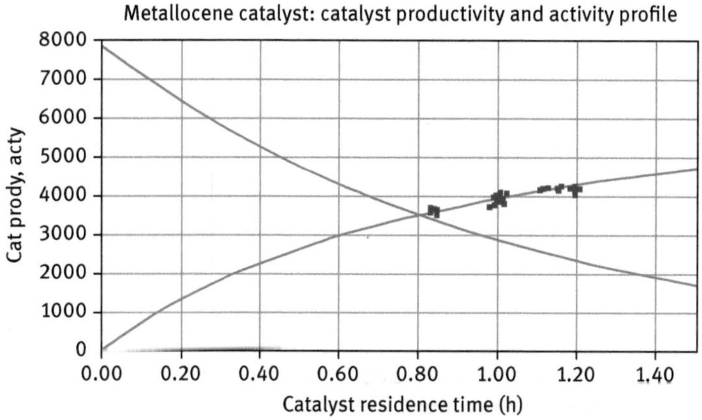

Fig. 8.2. Experimental Catalyst Productivity v. Residence Time curve and derived Catalyst Activity Profile for a Metallocene Catalyst; a=1.000; b=7874.

concentration may be generated. The strong decline of activity is, however, at the same time offering a solution to avoid the risk. Because of the rapid loss of activity, there is almost no advantage in seeking long residence times by striving for high solids concentrations. So the risk of operation is reduced by running the loop at lower solids concentrations, without sacrificing much of the catalyst productivity.

Figure 8.3 shows results obtained for a CrOx catalyst. The wider variation of the data points relative to the fitted curve is due to inaccuracies in the catalyst feed measurement. The constants are $a = 0.136$ and $b = 5065$. The activity of a CrOx catalyst starts at a rather low initial value, but is not declining so much with lifetime. The in-

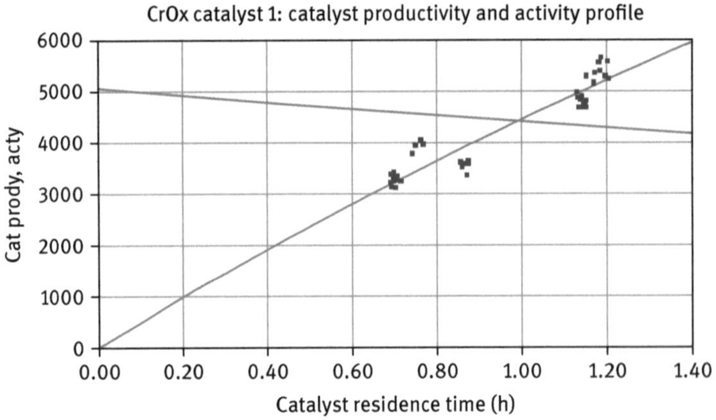

Fig. 8.3. Experimental Catalyst Productivity v. Residence Time curve and derived Catalyst Activity Profile for CrOx catalyst nr. 1; a=0.136; b=5065.

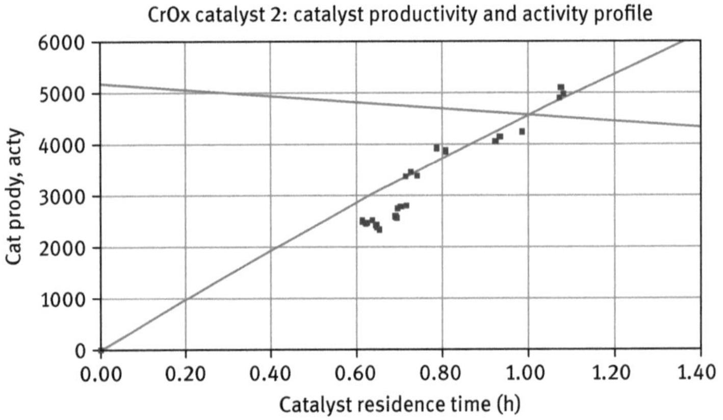

Fig. 8.4. Experimental Catalyst Productivity v. Residence Time curve and derived Catalyst Activity Profile for CrOx catalyst nr. 2; a=0.125; b=5173.

crease of catalyst productivity with residence time is consequently stronger than for the other catalyst families. So, for reasons of economy, it is important to run CrOx catalysts at the highest solids concentrations and long residence times, lest more of the already low activity is lost due to short residence times.

Figure 8.4 shows results obtained for another CrOx catalyst. The constants are $a = 0.125$ and $b = 5173$. The value of a is very close to the value found with the former CrOx catalyst, and is probably rather general for those catalysts. The similar values for b are more of a coincidence, because the first CrOx catalyst was run at a low activation temperature and a high hexene concentration, whilst this one was run

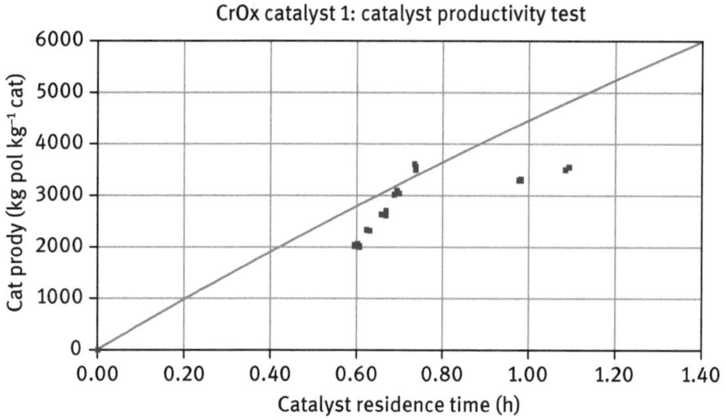

Fig. 8.5. Experimental Catalyst Productivity v. Residence Time curve for CrOx catalyst nr. 1. This test failed to yield a Catalyst Activity Profile. The Productivity curve shown is the one obtained from the same catalyst in a second experiment, Figure 8.3.

without hexene, but at a high activation temperature. In both cases the mentioned parameters affect the catalyst activities in opposite directions, resulting in more or less equal activities.
The two groups of data points at the shortest residence times fall below the fitted curve, and have not been taken into account for the curve fitting. It seems that at shorter residence times, or higher production rates, a heavy loss in catalyst activity is incurred. The reason behind this is discussed in Chapter 10, Section 10.6.

Figure 8.5 is illustrating that tests are not always successful. This test was a first attempt to record a productivity curve for the first CrOx catalyst. The results do not allow to draw any conclusion. The productivity curve that is shown for reference was copied later from the second, successful, test with the same catalyst.
The large group of data points at the shorter residence times fall below the productivity curve, and coincide with the same kind of data points in Figure 8.4, which is discussed in Chapter 10, Section 10.6.
The data points at higher residence times, around 1.0 and 1.1 *h*, are also very much below expectations. It was found that at those conditions the frequency of the catalyst injections from the discontinuous catalyst feed system, was in phase with the frequency at which the slurry is running the loop. The consequence is that the catalyst is not distributed evenly along the loop, and that zones of high catalyst concentrations are created, so-called *hot spots*. In those hot spots a shortage of ethylene is most probably limiting the catalyst productivity. How to avoid hot spots has been explained in Chapter 2, Section 2.3.3.

8.5 Catalyst Activity Profiles Showing an Induction Time

It is often stated that the catalyst activity only starts off some minutes after the first contact with the reactants. This time delay is called the *induction time.*

In that case, if the activity starts at time $\theta = \theta_0$ instead of $\theta = 0$, the induction time is equal to θ_0, and the activity profile is shifted along the time axis from 0 to θ_0,

$$\begin{aligned} \theta < \theta_0 : \quad & \overline{\mathcal{A}}(\theta, Rx) = 0 \,, \\ \theta \geq \theta_0 : \quad & \overline{\mathcal{A}}(\theta, Rx) = b\mathrm{e}^{-a(\theta-\theta_0)} \,. \end{aligned}$$

If a function $g(t) = f(t-t_0)$ for $t \geq t_0$, and $g(t) = 0$ for $t < t_0$, then the second translation property of the Laplace transform reads,

$$\mathcal{L}\{g(t)\} = \mathrm{e}^{-pt_0}\mathcal{L}\{f(t)\} \,.$$

So, according to (8.1), the productivity curve in case of an induction time is equal to,

$$\mathbb{P}(\tau, Rx) = \mathcal{L}\{\overline{\mathcal{A}}(\theta, Rx)\} = \mathrm{e}^{-\theta_0/\tau} b \frac{\tau}{1 + a\tau} \,. \tag{8.7}$$

When fitting this type of productivity curve to the data points from the different experiments, described in the previous section, the results have always been worse than for the case without induction time. So it is concluded that there is no need to take into account an induction time in order to represent full scale loop reactor activity profiles.

Notations

a	constant in the expressions for the catalyst activity profile and catalyst productivity, (s^{-1}).
$\mathbb{A}$	catalyst reactor activity: quantity of polymer produced by all catalyst present in the reactor per unit time, per *kg* catalyst, (*kg* polymer/*kg* catalyst, *s*).
$\overline{\mathcal{A}}(t, Rx)$	average catalyst activity over all catalyst particle sizes (or any other catalyst property): quantity of polymer produced per *kg* of catalyst, per unit time, at a certain lifetime *t* in the reactor, (*kg* polymer/*kg* catalyst, *s*).
b	constant in the expressions for the catalyst activity profile and catalyst productivity, (kg, kg^{-1}, s^{-1}).
$\mathbb{P}$	catalyst reactor productivity: total quantity of polymer produced by one *kg* of catalyst that has left the reactor, (*kg* polymer/*kg* catalyst).
$\mathcal{P}$	catalyst particle productivity: quantity of polymer produced by one catalyst particle during its lifetime at a certain set of reactor conditions, (*kg* polymer/*kg* catalyst).

Rx	symbol representing all of the reactor conditions, including temperature, concentrations, and flows.
t	catalyst lifetime: the time interval between the moment when a catalyst particle that is present in the reactor, has entered the reactor, and any moment at which the reactor system is considered, (*s*).
θ	catalyst residence time: the total time that a catalyst particle has spent in the reactor at the moment when it leaves the reactor, (*s*).
θ_0	induction time: the time delay at which the catalyst activity, after the first contact with the reactants, starts, (*s*).
τ	average catalyst residence time, (*s*).

9 Conversions of the Reactants

The catalyst reactor activity 𝔸 has been introduced in Chapter 5 without specifying that up to three reactants, viz. ethylene, hydrogen and 1-hexene (or another comonomer), participate in the polymerisation. Actually, the catalyst activity is the sum of the three activities from the three reactants. Those three activities are absolute quantities, which are quite different in magnitude.
An alternative way to describe the reaction of the three reactants, is to make use of conversions, which are more easily obtained from full scale reactor data, and are more convenient in calculations. The *conversion of reactant i*, ξ_i, is equal to the fraction of the reactant i that is consumed in the polymerisation reactor relative to the quantity of that reactant that is fed to the reactor.
As conversions are relative quantities, they do not offer sufficient information to calculate absolute quantities like catalyst activity and catalyst productivity. But they do allow to calculate the fraction of each reactant that is incorporated in the polymer. Those fractions are closely linked to quality parameters like melt index and density, and that is what makes conversions so important to the quality control in the reactor.

Conversions depend on the catalyst type and the reactor conditions, but also on the type of reactor and its geometry. Fortunately, it seems that the conversions in the loop reactor are little dependent on the production rate, and that greatly simplifies the control of the quality in that reactor.

9.1 Calculating Conversions from Full Scale Reactor Data

In this section, it is explained how the conversion of any of the three reactants is calculated, presenting the ethylene conversion as an example. The method, however, is valid for all of the three reactants, and the formulas are readily adapted by the simple replacement of an index and a molecular weight.

The most obvious calculation of the ethylene conversion, under steady state conditions, is derived from a simple ethylene mass balance for the reactor, stating that the ethylene that is not consumed, is found in the flash gas,

$$\xi_2 = 1 - \frac{D_R\, y_2}{F_2}\frac{28.05}{(\mathrm{MW})_{\mathrm{fg}}} = 1 - \frac{(\sum F_i - P) y_2}{F_2}\frac{28.05}{(\mathrm{MW})_{\mathrm{fg}}}, \tag{9.1}$$

where ξ_2 is the conversion of ethylene (-), F_2 is the ethylene feed to the reactor (kg, s^{-1}), D_R is the flash gas flow from the reactor (kg, s^{-1}), $\sum F_i$ is the sum of all feeds to the reactor (kg, s^{-1}), P is the production rate of polymer (kg, s^{-1}), and y_2 is the ethylene concentration in the flash gas (*mole fraction*).

Equation (9.1) yields quite accurate results here, since the quantities and concentrations of ethylene are quite large. However, it cannot be used for the comonomer and hydrogen, due to a great lack of accuracy, because there the flows and concentrations are much smaller. Therefore, the following calculation of the conversion is highly recommended. It is the only method that yields sufficient accuracy when calculating the conversions of hydrogen and the comonomer. It also allows to screen the data points more easily for stability, which is important to obtain reliable data.

The same mass balance (9.1) for ethylene is written in a slightly different form,

$$F_2(1-\xi_2) = D_R\, y_2 \frac{28.05}{(\mathrm{MW})_{\mathrm{fg}}}\,. \tag{9.2}$$

An analogous mass balance can be written for the diluent, here in terms of an isobutane diluent,

$$F_{\mathrm{IC4}}(1-\xi_{\mathrm{IC4}}) = D_R\, y_{\mathrm{IC4}} \frac{58.12}{(\mathrm{MW})_{\mathrm{fg}}}\,.$$

Since isobutane is inert in this reaction, its conversion ξ_{IC4} is zero, or, all isobutane ends up in the flash gas,

$$F_{\mathrm{IC4}} = D_R\, y_{\mathrm{IC4}} \frac{58.12}{(\mathrm{MW})_{\mathrm{fg}}}\,. \tag{9.3}$$

Dividing equation (9.2) by equation (9.3),

$$\frac{F_2(1-\xi_2)}{F_{\mathrm{IC4}}} = \frac{y_2}{y_{\mathrm{IC4}}}\frac{28.05}{58.12}\,,$$

whence the recommended expression for the ethylene conversion is obtained,

$$\xi_2 = 1 - \frac{F_{\mathrm{IC4}}}{F_2}\frac{y_2}{y_{\mathrm{IC4}}}\frac{28.05}{58.12}\,. \tag{9.4}$$

This expression does not include the flash gas flow, and is based on proportions instead of absolute values, which makes it more accurate.
A great restriction to its use is that it is only valid for the steady state, at very stable conditions. It must not be used for real time calculations, lest it would produce nonsense and leave the reactor operator in the greatest confusion.

The practical way to calculate conversions from historical data, is to collect a number of data points based on hourly averages of the process parameters, and tabulated in chronological order. Those data points are plotted, see Figure 9.1, with the ratios of ethylene to isobutane in the reactor feed, F_2/F_{IC4}, as values of the abscissa, and the molar ratios of ethylene to isobutane in the flash gas, y_2/y_{IC4}, as the ordinate values. The origin (0, 0) is included in the plot. A straight line is drawn from the origin to a data point, or to the centre of a cluster of data points. The slope of this line,

$$(\mathrm{slope}) = \frac{y_2/y_{\mathrm{IC4}}}{F_2/F_{\mathrm{IC4}}} = \frac{F_{\mathrm{IC4}}}{F_2}\frac{y_2}{y_{\mathrm{IC4}}}\,,$$

is related to the conversion, according to equation (9.4),

$$\xi_2 = 100 - 100(\mathrm{slope})\frac{28.05}{58.12}\,,$$

where practical units are % for ξ_2, *tonne* per *tonne* for the abscissa F_2/F_{IC4}, and *mole* per *mole* for the ordinate y_2/y_{IC4}.

Conversions seem to be quite independent of the production rate. Data points taken from identical reactor conditions, but at different production rates, are located on one line to the origin. When reactor conditions are changed, the data points will move to a different line. This property is used to screen the plotted data points carefully for stability.

One line, or a small number of lines, is drawn from the origin to the clusters of data points, for reference. The data points are then inspected on the plot in chronological order. (Commercial spreadsheet programs do allow to move a cursor on a plot from point to point in the order of tabulation.) Data points from identical reactor conditions will stick close to such a reference line. In unstable periods, the points make larger jumps on the plot from one hour to the other, or follow a trajectory at large pace from one cluster to the other. Those unstable points are rigorously removed from the data set. The ethylene conversions are calculated for the remaining stable data points. They are used to establish correlations, characterise or compare catalysts, and build models to serve quality control.

Examples of such plots are shown in Figures 9.1, 9.3, and 9.5. The lower the conversion is, the steeper the slope of the line. Indeed, more ethylene will end in the flash gas at lower conversions.

Even if real time use of equation (9.4) is absolutely not recommended, it can support the reactor operation more directly, e.g. when a certain step should be taken to make a larger adjustment to melt index or density, or when a grade change is prepared. Then a period of at least three consecutive hours of the current stable operation should be selected. The stability of these data points is a requisite. The average values, of the hourly averages, of the process parameters from that period are calculated and used as input in equation (9.4). The obtained ethylene conversion, together with the other conversions, is used to estimate the necessary step to be taken, with the help of an off-line model. Such a model is proposed later in this book, Chapter 11.

9.2 The Conversion of Ethylene

At first sight, the conversion of ethylene is quite constant at around 95%, and often very little importance is attached to it. However, small variations in the ethylene conversion effect a great influence on the ratios of hydrogen and 1-hexene to ethylene in the reactor, and thus on the quality of the product, even on the stability of the reactor operation. Since the conversion of ethylene is most strongly influenced by the ethylene concentration itself, one of the important rules of operation of an HDPE slurry loop reactor is to keep the ethylene concentration, and thus the ethylene conversion, well under control and as constant as possible.

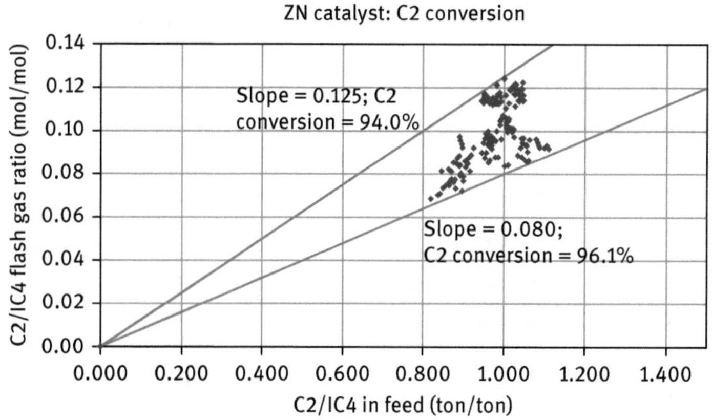

Fig. 9.1. Ethylene to isobutane ratios in the reactor feeds and in the flash gas, ZN catalyst. The slopes of the lines from the origin to the data points are related to the ethylene conversions.

The calculation of the ethylene conversion is based on equation (9.4),

$$\xi_2 = 1 - \frac{F_{IC4}}{F_2}\frac{y_2}{y_{IC4}}\frac{28.05}{58.12},$$

and effected as described in Section 9.1, making use of

$$\xi_2 = 100 - 100(\text{slope})\frac{28.05}{58.12},$$

where practical units are % for ξ_2, *tonne* per *tonne* for the abscissa F_2/F_{IC4}, and *mole* per *mole* for the ordinate y_2/y_{IC4}. Stable data points from a long run with a Ziegler–Natta catalyst, at widely different conditions, are shown in Figure 9.1. The two lines to the origin show the limits in between which the ethylene conversion did vary. The lower line indicates a conversion of 96.1%, whilst the upper line is at 94.0%. A such difference of 2% in conversion is of great consequence to the reactor operation, as will be explained in the next two sections.

The ethylene conversion can be correlated with the reactor conditions, and is found to be independent of the production rate.

For one *Ziegler–Natta catalyst*, a correlation of the following type was established by linear regression,

$$\xi_2 = \beta_0 + \beta_1 y_2 + \beta_2 y_{cocat} + \beta_3 T .$$

The ethylene conversion varies from 94 to 97%. The correlation is explaining almost 95% of the total variance, of which some 84% is taken into account by the ethylene concentration y_2, obviously the greatest factor. Its coefficient β_1 is negative, which seems to be acceptable, as, when more ethylene is present in the reactor, a larger fraction may leave unreacted. The coefficients for the cocatalyst, triethylaluminium (TEAl), concentration y_{cocat}, and the temperature T are positive and seem to indicate a slightly higher activity of the catalyst at higher cocatalyst concentration and higher temperature.

The following type of correlation was established based on the massed data from three different *CrOx catalysts*,

$$\xi_2 = \beta_0 + \beta_1(\text{cat1}) + \beta_2(\text{cat2}) + \beta_3 y_2 + \beta_4 T\ .$$

The ethylene conversion varies from 95 to 98%. The correlation is explaining 80% of the total variance. Some 67% is taken into account by the ethylene concentration y_2, also here the greatest factor. Its coefficient β_3 is negative, like it is for the Ziegler–Natta catalyst.
CrOx catalysts can be used either pure or in blends. The factors (cat1) and (cat2) represent the fractions of two of the three CrOx catalysts that were used in the blends. The fraction of the third catalyst is of course determined by the former two, (cat3) = 1 – (cat1) – (cat2). The effects of reactor conditions like temperatures and concentrations, β_3, β_4, on the catalyst behaviour are the same for all CrOx catalysts. This offers the great advantage that data from different CrOx catalysts can be treated in one set of data, and yield more accurate correlations. But each catalyst's response is situated at another level. This level is specified by the factors (cat1) and (cat2). These are not just "dummy variables", with discrete values of either zero or one, but continuous variables from zero to one, equal to the fraction of each catalyst used in the blends.
The coefficient for the temperature T is negative, which is difficult to explain, and maybe doubtful. However the effect of the temperature is of a minor order.

For one Zr based *metallocene catalyst*, the ethylene conversion shows little variation around an average of 95.4%. A dependency on the ethylene and 1-hexene concentrations was found by linear regression,

$$\xi_2 = \beta_0 + \beta_1 y_2 + \beta_2 y_6\ .$$

The correlation is explaining only 49% of the total variance, but, as mentioned, the variance is already small. The coefficients β_1 and β_2 are negative, as for the other catalysts. There is no effect from the temperature T, even if the data was taken from a considerable range in temperature, from 85 to 95 °C.

A note on the dosage of cocatalysts
In the correlations that are given for the conversions, the cocatalyst concentration y_{cocat} is expressed and controlled as a number of weight *ppm* in the diluent (isobutane) feed flow to the reactor.
Cocatalysts are often dosed in a constant ratio to the catalyst feed, like a constant Al:Ti ratio for TEAl with a Ziegler–Natta catalyst, or a constant B:Cr ratio for triethylboron (TEB) with CrOx catalysts. This tradition stems from lab scale polymerisations, where chemists dose it that way.
In full scale polymerisation, it is recommended to dose them as weight *ppm* in the diluent. The reasoning behind it is that the environment met by the catalyst in the reactor must be determining its behaviour. Imagine a tiny catalyst particle in a huge pool of reactor liquid. That particle will have no idea of the Al:Ti or B:Cr ratio, but it will face the cocatalyst concentration.

When maintaining a constant Al:Ti or B:Cr ratio in the reactor, the TEAl or TEB concentrations will change all the time when the liquid and solids concentrations or the catalyst productivity vary. This may even create risky situations. E.g., when the catalyst activity falls for some reason, like the presence of a catalyst poison, keeping up the production rate will require a higher catalyst feed rate. In the case of a CrOx catalyst with TEB cocatalyst, maintaining a constant B:Cr ratio will then lead to a higher TEB feed rate and concentration. TEB is generating 1-hexene and thus lowering the density. When feeding more TEB the polymer density falls, and the fouling curve may be crossed.
Besides a such incident, the experience that the cocatalyst concentration, expressed as *wtppm* in the diluent, has a significant and reproducible effect on the conversions of the reactants and on the catalyst productivity, has confirmed the philosophy here described.

9.3 The Conversion of Hydrogen

Hydrogen has the smallest molecular weight of all chemical substances, and so its flow in weight units may be disregarded in most of the mass balances around the reactor. Still, it is a reactant just like the other reactants, and its conversion can be calculated. And even if it is hardly traceable in the polymer, every hydrogen molecule that reacts will terminate a polymer chain, and thus affect the average molecular weight of the polymer, or its melt index. This makes the conversion of hydrogen and its incorporation in the polymer a very important parameter. When catalysts, like metallocenes and Ziegler–Natta, are said to have a high *hydrogen response*, which means that the melt index is strongly increasing with the hydrogen concentration, this is explained by their high conversion of hydrogen. The hydrogen to ethylene feed ratio will be used to control the melt index with these catalysts. A low conversion of hydrogen is found with catalysts that show a low hydrogen response, like CrOx catalysts, where the hydrogen will not be used to control the melt index.

9.3.1 Hydrogen to Ethylene Ratios

The ratio of hydrogen to ethylene in the flash gas is determined by the conversions of both hydrogen and ethylene. Writing a mass balance for hydrogen, analogous to the one for ethylene (9.2),

$$F_{\mathrm{H}}(1 - \xi_{\mathrm{H}}) = D_{\mathrm{R}}\, y_{\mathrm{H}} \frac{2.01}{(\mathrm{MW})_{\mathrm{fg}}}\,, \tag{9.5}$$

and dividing it by the same mass balance (9.2) for ethylene,

$$F_2(1 - \xi_2) = D_{\mathrm{R}}\, y_2 \frac{28.05}{(\mathrm{MW})_{\mathrm{fg}}}\,,$$

is obtained,

$$\frac{F_{\mathrm{H}}}{F_2}\frac{1 - \xi_{\mathrm{H}}}{1 - \xi_2} = \frac{y_{\mathrm{H}}}{y_2}\frac{2.01}{28.05}\,. \tag{9.6}$$

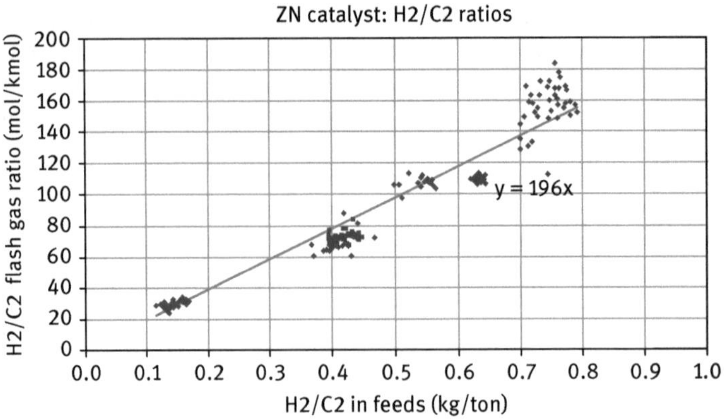

Fig. 9.2. Hydrogen to ethylene ratios in the reactor feeds and in the flash gas. ZN catalyst.

This equation (9.6) explains the experience of the reactor operator that the hydrogen to ethylene (H_2/C_2) ratio in the flash gas will rise when the (H_2/C_2) ratio in the feeds is increased, see Figure 9.2. It also shows how the (H_2/C_2) ratio in the flash gas will be higher at equal ratios in the feeds when the hydrogen conversion is lower, since more of the fed hydrogen will leave as part of the flash gas.

Important to note is that the (H_2/C_2) ratio in the flash gas is also determined by the conversion of ethylene, which is mainly determined by the ethylene concentration. Consequently, it is not wise to specify a (H_2/C_2) ratio in the flash gas in a product recipe without specifying the ethylene concentration, because a change in ethylene concentration will change the ethylene conversion, and thus the required (H_2/C_2) ratio in the flash gas. The strong effect of the ethylene conversion becomes here evident: a change in conversion from 94 to 97% will boost the (H_2/C_2) ratio in the flash gas by a factor $(1 - 0.94)/(1 - 0.97) = 2$, i.e. double it.

9.3.2 Calculating the Hydrogen Conversion

The calculation of the hydrogen conversion is based on the following equation obtained as shown in Section 9.1,

$$\xi_H = 1 - \frac{F_{IC4}}{F_H} \frac{y_H}{y_{IC4}} \frac{2.01}{58.12}, \tag{9.7}$$

and effected as described in the same Section 9.1, using the practical expression,

$$\xi_H = 100 - 100(\text{slope})\frac{2.01}{58.12},$$

where units are % for ξ_H, *kg* H_2 per *tonne* of isobutane for the abscissa F_2/F_{IC4}, and *mole* of H_2 per *kilomole* of isobutane for the ordinate y_H/y_{IC4}.

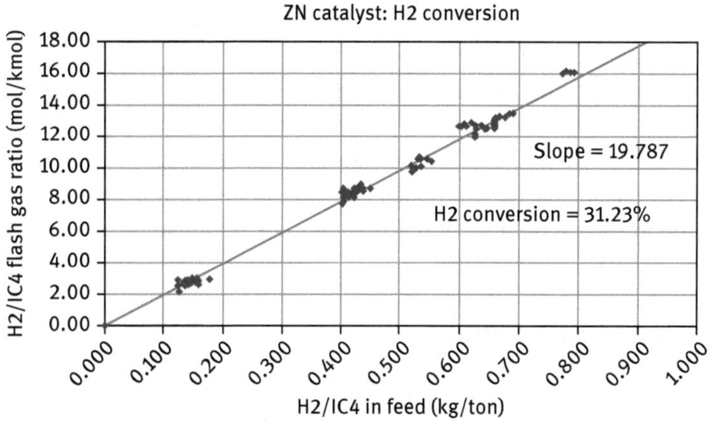

Fig. 9.3. Hydrogen to isobutane ratios in the reactor feeds and in the flash gas, ZN catalyst. The slopes of the lines from the origin to the data points are related to the hydrogen conversions.

Figure 9.3 is showing an example. The data points are the same stable data points as shown in the previous Figure 9.2. They are clearly less scattered when plotted as ratios to the diluent, instead of to ethylene, because the effects of changing ethylene conversions are not visible. The lower the conversion is, the steeper the slope of the line.

For the hydrogen conversion by one *Ziegler–Natta catalyst*, a correlation of the following type was established by linear regression,

$$\xi_H = \beta_0 + \beta_1 y_2 + \beta_2 (y_2)^2 + \beta_3 y_{cocat} + \beta_4 y_H \,.$$

The hydrogen conversion varies from 28 to 38%. The correlation is explaining almost 78% of the total variance, of which some 55% is taken into account by the ethylene concentration y_2, again the greatest factor. A quadratic term is needed to reflect its full impact. Its coefficient β_1 is negative, which, though this kind of explanations is always risky, seems acceptable, as, when more ethylene is present in the reactor, it will be more difficult for the hydrogen molecules to compete with the ethylene in the diffusion race to the catalyst surface. The coefficient for the cocatalyst (TEAl) concentration y_{cocat} is positive and explains another 18% of the variance. There is no effect from the temperature. An additional small positive effect from the hydrogen concentration y_H itself explains some 4% of the variance. The latter can be omitted in models without much loss of accuracy to avoid iterative calculations.

The hydrogen conversion with *CrOx catalysts* is very low at around 4%. This explains why hydrogen is not used to control the melt index with CrOx catalysts in the first place. The melt index is there much more sensitive to the reactor temperature and the catalyst activation temperature. Very often no hydrogen is used with CrOx catalysts at all.

At such low conversions even the expression (9.7) does not yield sufficient accuracy to avoid negative and thus impossible values. A meaningful correlation to the reactor conditions could not be found. The best results are obtained for CrOx catalysts by postulating a constant hydrogen conversion at 3.8%.

For a *metallocene catalyst*, the hydrogen conversion is also constant, though at the other end of the scale, and simply equal to 100% ! This is not just an approximation. When establishing models for metallocene catalysts, the readings of the hydrogen concentration in the flash gas are just noise, and should be set at a simple zero. The hydrogen conversion should be set at 100% to obtain the best results.

9.4 The Conversion of 1-Hexene

A comonomer like 1-hexene is used to control the density of the polymer. The more comonomer is incorporated in the polymer, the lower the density will be, due to the short sidechain added to the linear polymer chain by the addition of each 1-hexene molecule.

By the catalyst's *comonomer response*, is meant how strongly the polymer density will decrease with the comonomer concentration in the reactor liquid. This is explained by the comonomer conversion and the corresponding amount of comonomer incorporated in the polymer.

The use of 1-hexene is described here, but also other α-olefins can be used as comonomers, like 1-butene, 1-octene, or 4-methyl-1-pentene, which is an isomer of 1-hexene. Apart from economical considerations, the selection of a comonomer will be based on the product quality (heavier comonomers yielding a stronger polymer), and the capability of the recovery section, downstream of the reactor, to separate the comonomer from the diluent. The latter will usually impose a comonomer that is heavier than the diluent.

It is important to have a catalyst with a high 1-hexene conversion in a loop reactor, because low conversions will make high concentrations of 1-hexene necessary to reach a certain density, and those high concentrations will bring the liquid composition closer to the fouling curve. So the 1-hexene conversion is setting one of the important limits to the operation of the slurry loop reactor. It is very difficult to reach lower densities with Ziegler–Natta catalysts, which show a low 1-hexene conversion, whilst metallocene catalysts that own high 1-hexene conversions allow even the production of linear low density (LLD) products of a density down to 920 $kg\,m^{-3}$. When selecting a Ziegler–Natta catalyst for the loop, the 1-hexene conversion is an important criterion, which may decide on the technical feasibility of certain grades.

9.4.1 1-Hexene to Ethylene Ratios

Just as for hydrogen, the ratio of 1-hexene to ethylene in the flash gas is determined by the conversions of both 1-hexene and ethylene. Dividing a mass balance for 1-hexene

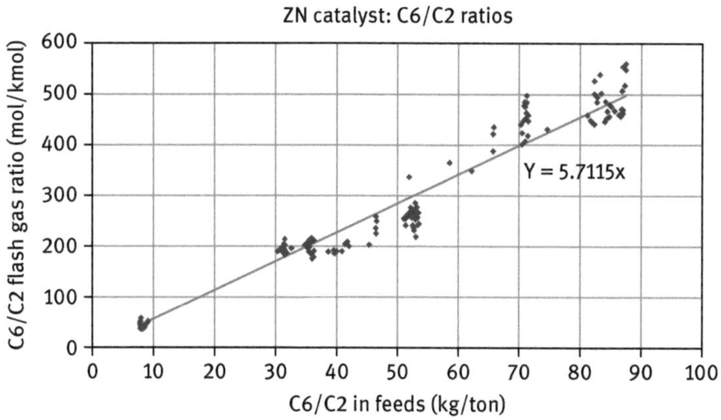

Fig. 9.4. 1-Hexene to ethylene ratios in the reactor feeds and in the flash gas. ZN catalyst.

by the mass balance (9.2) for ethylene, an analogous expression is obtained as for the hydrogen to ethylene ratio in Section 9.3.1,

$$\frac{F_6}{F_2}\frac{1-\xi_6}{1-\xi_2}=\frac{y_6}{y_2}\frac{84.16}{28.05}\,. \tag{9.8}$$

This equation (9.8) explains the experience of the reactor operator that the 1-hexene to ethylene (C_6/C_2) ratio in the flash gas will rise when the (C_6/C_2) ratio in the feeds is increased, see Figure 9.4. It also shows how the (C_6/C_2) ratio in the flash gas will be higher at equal ratios in the feeds when the 1-hexene conversion is lower, since more of the fed 1-hexene will leave as part of the flash gas.

Also here it is important to note that the (C_6/C_2) ratio in the flash gas is also determined by the conversion of ethylene, which mainly depends on the ethylene concentration. So it is not wise either to specify a (C_6/C_2) ratio in the flash gas in a product recipe without specifying the ethylene concentration. The strong effect of the ethylene conversion is here very important. A change in conversion from 94 to 97% will boost the (C_6/C_2) ratio in the flash gas by a factor $(1-0.94)/(1-0.97)=2$, i.e. double it. This means that, when a producer wants to run a product of a lower density, especially with a Ziegler–Natta catalyst, it is crucial to select a set of reactor conditions where the conversion of 1-hexene is maximal in relation to the conversion of ethylene. This will allow to incorporate the highest proportion of 1-hexene to ethylene, at the lowest concentration of 1-hexene. This choice may decide whether or not such a product is technically feasible in the loop reactor, as the liquid composition may lie at the one or the other side of the fouling curve. On this topic is more elaborated in Section 9.7.

9.4.2 Calculating the 1-Hexene Conversion

The calculation of the 1-hexene conversion is based on the same equation obtained in Section 9.1,

$$\xi_6 = 1 - \frac{F_{\text{IC4}}}{F_6}\frac{y_6}{y_{\text{IC4}}}\frac{84.16}{58.12}\,, \tag{9.9}$$

and effected as described in the same Section 9.1, using the practical expression,

$$\xi_6 = 100 - 100(\text{slope})\frac{84.16}{58.12}\,,$$

where practical units are % for ξ_6, *kg* of 1-hexene per *tonne* of isobutane for the abscissa F_6/F_{IC4}, and *mole* of 1-hexene per *kilomole* of isobutane for the ordinate y_6/y_{IC4}. Figure 9.5 is showing an example. The data points are the same stable data points as shown in the previous Figure 9.4. They are clearly less scattered when plotted as ratios to the diluent, instead of to ethylene, because they do not reflect the effects of the changing ethylene conversions.

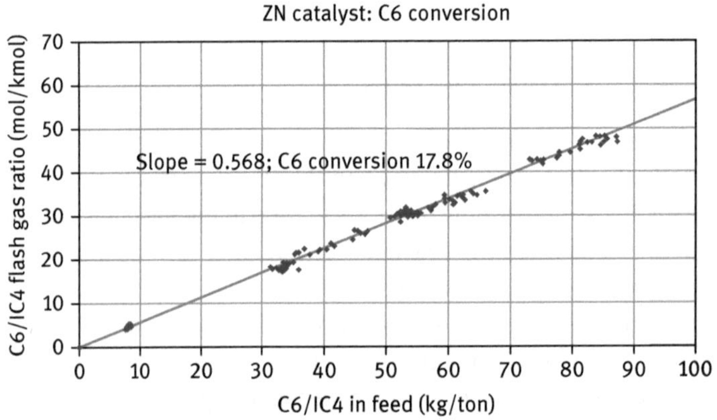

Fig. 9.5. 1-Hexene to isobutane ratios in the reactor feeds and in the flash gas. ZN catalyst. The slopes of the lines from the origin to the data points are related to the 1-hexene conversions.

For the 1-hexene conversion by the same *Ziegler–Natta catalyst*, a correlation of the following type was established by linear regression,

$$\xi_6 = \beta_0 + \beta_1 y_2 + \beta_2 y_{\text{cocat}} + \beta_3 T\,.$$

The correlation is explaining 65% of the total variance, of which almost 50% is taken into account by the combined actions of TEAl, y_{cocat}, and temperature, T, whilst the ethylene concentration is explaining around 24%. The coefficient β_1 for the ethylene concentration is negative, like has been found for the ethylene and hydrogen conversions. The coefficients for the cocatalyst (TEAl) concentration and the temperature are positive again.

The 1-hexene conversion varies from 12 to 25%. These low values necessitate high 1-hexene concentrations. Fortunately, relatively low reactor temperatures are possible and the rather narrow molecular weight distribution includes less low molecular weight, so that the fouling curve can be avoided.

The 1-hexene conversion with *CrOx catalysts* is around 70%. Low densities can be reached, down to some 935 $kg\,m^{-3}$, the area of the so called medium density HDPE's. But this range of conversion, together with the required reactor temperatures, and the broad molecular weight distribution, which always contains a tail of lower molecular weight, makes the fouling curve of a constant concern to the operator.
A correlation for the 1-hexene conversion by two different CrOx catalysts, when no cocatalyst like TEB is used, shows a dependency on the ethylene concentration and on the catalyst,

$$\xi_6 = \beta_0 + \beta_1(\text{cat1}) + \beta_2 y_2 \,.$$

The 1-hexene conversion varies from 65 to 80%. The correlation is explaining 88% of the total variance. The coefficient of the ethylene concentration, β_2, is negative and consistent with all other correlations.
The factor (cat1) represents the fraction of one of the two CrOx catalysts used in the blends. As explained above when discussing the ethylene conversion, the effects of reactor conditions on the catalyst behaviour are the same for all CrOx catalysts, whilst each catalyst's response is situated at another level, here specified by the factor (cat1). The data was too limited to quantify an effect from the reactor temperature, or activation temperature.

For a *metallocene catalyst*, the 1-hexene conversion may be as high as 70%. This high conversion, together with a very narrow molecular weight distribution, allows to produce very low densities, down to LLD products. To illustrate this effect, incorporating the same fraction of 1-hexene into the polymer with a metallocene catalyst at 70% conversion, compared to a Ziegler–Natta catalyst at 20% conversion, will take a 1-hexene concentration of $(1-0.20)/(1-0.70) = 2.66$, or 2.7 times less.
For the same Zr based metallocene catalyst, mentioned in Section 9.2, the 1-hexene conversion is lower, and shows little variation around an average of 57%. A dependency on the ethylene and 1-hexene concentrations is explaining 45% of the total variance,

$$\xi_6 = \beta_0 + \beta_1 y_2 + \beta_2 y_6 \,.$$

This correlation is analogous to the correlation for the ethylene conversion by the same catalyst. The coefficients β_1 and β_2 are negative again, and there is no effect from the temperature T in the broad temperature range, between 85 and 95 °C.

The 1-hexene conversion for *chromocene catalyst* is zero! This feature makes this catalyst very interesting to be combined with a CrOx catalyst in the loop reactor, and produce bimodal products. The density of the product from the CrOx catalyst can be controlled independently, as only that catalyst will incorporate 1-hexene. On the other hand, the high sensitivity of the chromocene catalyst to hydrogen, to which the CrOx hydrogen sensitivity vanishes, allows an independent control of the melt index of the chromocene product.

9.5 Conversions and Polymer Properties

Most important to produce a suitable HDPE is the careful selection of the catalyst. The quality of an HDPE is indicated in product data sheets by its melt index (MFI) and density (DENS). However, an infinite number of products can be imagined showing the same values of MFI and DENS. A true definition of an individual product should include the molecular weight distribution, the type and amount of incorporated comonomer, and the distribution of the incorporated comonomer over the molecular weight, or chain length. Most of this is achieved by the proper selection and preparation of the catalyst. All that is left to the reactor operator is to control the MFI and DENS at the prescribed level. But even the range wherein MFI and DENS can be controlled, is defined by the catalyst.
The degree of branching is another aspect of the quality. HDPE produced in a slurry loop reactor is linear, i.e. without long branches to the chain. Some long chain branching (LCB) occurs in the extruder by the combined action of mechanical energy, temperature and oxygen.

MFI and DENS are usually controlled in the reactor by ratios of concentrations, 1-hexene and hydrogen to ethylene, in the reactor liquid. This is highly imprecise, as was shown in this chapter, because the ratios of concentrations depend on the conversions of the reactants, and those are not constant. The best control is obtained by relating MFI and DENS as directly as possible to the composition of the polymer, and that is, for a given catalyst system, to the *fraction of incorporated hydrogen*, x_H, and the *fraction of incorporated 1-hexene*, x_6. Those two fractions are known from the conversions,

$$x_\mathrm{H} = \frac{F_\mathrm{H}\xi_\mathrm{H}}{P}\,, \tag{9.10}$$

$$x_6 = \frac{F_6\xi_6}{P}\,. \tag{9.11}$$

Correlations for MFI and DENS, produced by the selected type of catalyst system, as a function of x_H and x_6, can be made from either lab scale data, or, preferably, from full scale reactor data, and used for control of MFI and DENS in the reactor.

9.5.1 Correlations for Melt Index

For a typical *Ziegler–Natta catalyst* the melt index is related to the incorporated hydrogen in the first place, and also to the incorporated 1-hexene and the reactor temperature,

$$\log(\text{MFI}) = \beta_0 + \beta_1 x_\text{H} + \beta_2 (x_\text{H})^2 + \beta_3 x_6 + \beta_4 T \,.$$

This correlation is explaining over 95% of the variance by the factor incorporated hydrogen, and 99.2% in total. A quadratic term in x_H is needed. Practical units are *wtppm* for x_H and *wt%* for x_6.
Note that MFI is always present in correlations as log(MFI).

For a *CrOx catalyst* the reactor temperature is the most important factor, together with the activation temperature T_act of the catalyst,

$$\log(\text{MFI}) = \beta_0 + \beta_1(\text{cat1}) + \beta_2(\text{cat2}) + \ldots + \beta_4 T + \beta_5 T_\text{act} + \beta_6 x_\text{H} + \beta_7 x_6 \,.$$

This correlation, covering results from 4 different catalysts, is explaining 96% of the total variance.
The factors (cat1), (cat2) ... represent the fractions of the CrOx catalysts used in the blends. The effects of reactor temperature, activation temperature, incorporated 1-hexene and hydrogen are positive, and common to all CrOx catalysts. Each catalyst's response is situated at another level, and specified by $(n-1)$ factors (cat1), (cat2), ..., when n catalysts are involved.

For the *metallocene catalyst* the melt index is only related to the incorporated hydrogen, without any effects from the incorporated 1-hexene or the reactor temperature,

$$\log(\text{MFI}) = \beta_0 + \beta_1 x_\text{H} + \beta_2 (x_\text{H})^2 \,.$$

This correlation is explaining over 97% of the total variance. The quadratic term in x_H cannot be neglected, and is similar to the Ziegler–Natta case. It confirms that these quadratic terms are not just a coincidence.
The total conversion of hydrogen and the fact that the melt index is determined by the incorporated hydrogen only, makes the melt index control in the loop reactor extremely stable and easy with metallocene catalysts. The only requisite is a precise dosing of low flows of hydrogen, because their high conversion makes the hydrogen feeds very small indeed.

A *chromocene catalyst* is also very sensitive to hydrogen. It is expected that the correlation for the melt index will be of the same type as for the metallocene catalyst.

9.5.2 Incorporated 1-Hexene and Density

The density is critical to the quality of HDPE. It affects mechanical properties like stiffness (elongational modulus), creep resistance (environmental stress cracking resistance) and impact strength. It should be measured and controlled at an accuracy of $0.1\ kg\,m^{-3}$.
HDPE is, like several other polymers (polypropylene, polyethylene terephthalate ...) semicrystalline in the solid state. It consists of a crystalline and an amorphous phase.

Each phase has its own constant density, and the resulting density of the product is a linear combination of those two densities, defined by the fraction of crystalline material in the polymer, or *crystallinity*.

The crystallinity, and thus the density, is a function of the incorporated 1-hexene, the melt index, and, not to forget, the sample's cooling history during solidification.

The effect of the incorporated 1-hexene is explained by the fact that each short side chain added to the polymer chain by the insertion of a comonomer molecule, so called *short chain branching* (SCB), interrupts the participation of the chain in the crystalline phase.

The density is increasing with a higher melt index, probably because short chains are more easily aligned and integrated in the crystalline phase.

The strong influence of the cooling history of the sample makes it necessary to follow a detailed procedure for the cooling of the sample from the molten phase, before the density is measured, which is described in the appropriate standard test methods.

A correlation for the density of an HDPE produced from *CrOx catalysts* is,

$$\text{DENS} = \beta_0 - \beta_1 (x_6)^{0.49} + 2.65 \log(\text{MFI}_{21}) . \quad (9.12)$$

This correlation, based on results from 3 different catalysts, is explaining 99.7% of the total variance. The melt index is measured at a weight of 21.6 *kg*. The unit for x_6 is *wt%*.

To represent the density as a sole function of the incorporated 1-hexene, the *reduced density* is defined as,

$$(\text{reduced density}) = \text{DENS} - 2.65 \log(\text{MFI}_{21}) ,$$

see Figure 9.6. This type of correlation, reduced density versus incorporated 1-hexene, is probably valid for all CrOx catalysts, within the covered range, down to a reduced density of 937. Extrapolation of this kind of correlations toward lower densities is risky, lower densities requiring larger and larger amounts of incorporated 1-hexene.

The effect of the melt index on the density of an HDPE produced from *Ziegler–Natta catalyst* is much stronger,

$$\text{DENS} = \beta_0 - \beta_1 (x_6)^{0.34} + 4.762 \log(\text{MFI}_2) .$$

The melt index is measured at a weight of 2.16 *kg*. The unit for x_6 is *wt%*. The explained variance is 97.3%.

The influence of the melt index on the density is so strong with Ziegler–Natta catalysts that it does not make sense to mention the density without specifying the melt index. The reduced density, defined as,

$$(\text{reduced density}) = \text{DENS} - 4.762 \log(\text{MFI}_2) ,$$

should be used instead, as it is directly linked to the required fraction of incorporated 1-hexene, see Figure 9.7. This type of correlation is less dependent on the type of

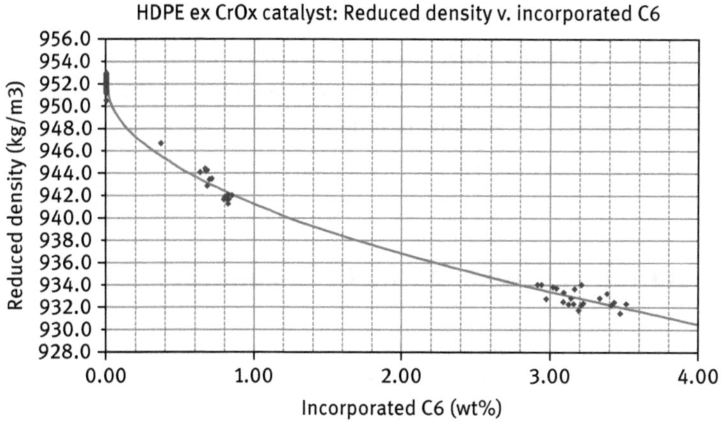

Fig. 9.6. Correlation for the reduced density versus incorporated 1-hexene for CrOx HDPE.

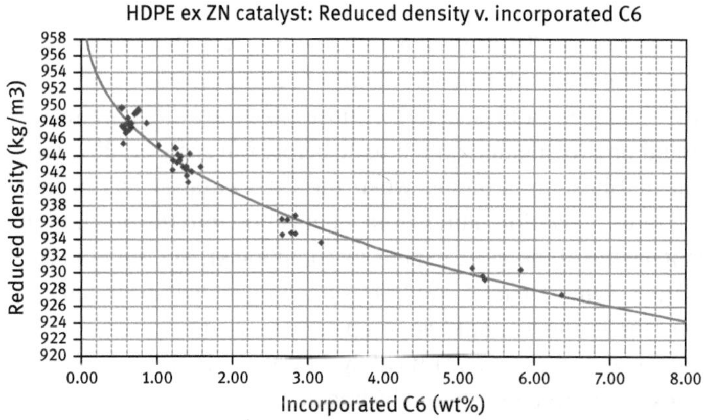

Fig. 9.7. Correlation for the reduced density versus incorporated 1-hexene for a Ziegler–Natta HDPE.

Ziegler–Natta catalyst, because the 1-hexene conversion, which differs from catalyst to catalyst, has already been taken into account, but the correlation is not general for all Ziegler–Natta catalysts. It was noted that two quite similar Ziegler–Natta catalysts produce different densities at the same fraction of incorporated comonomer, with density differences up to 2 units. This might be explained by a difference in the distribution of the comonomer in the polymer chains.

More and more incorporated 1-hexene is required to reach lower densities. Due to the low 1-hexene conversions for a Ziegler–Natta catalyst, the required 1-hexene concentrations for the lower densities are extremely high.

The density obtained from the *metallocene catalyst* follows the same general type of correlation,

$$\mathrm{DENS} = \beta_0 - \beta_1 (x_6)^{0.45} + 4.511 \log(\mathrm{MFI}_2)\,.$$

The melt index is measured at a weight of 2.16 *kg*. The unit for x_6 is *wt%*. The explained variance is 99.7% in a density range from 921 to 962 $kg\,m^{-3}$.
The effect of the melt index on the density is strong and very close to the one found for Ziegler–Natta catalysts. The reduced density is here,

$$(\text{reduced density}) = \text{DENS} - 4.511 \log(\text{MFI}_2)\,.$$

9.6 In-situ Generation of 1-Hexene with CrOx Catalysts

The presence of triethylboron (TEB) as a cocatalyst near the CrOx catalyst surface, (effected by dosing a solution of liquid TEB in the diluent, at an order of magnitude of 1 *wtppm* of TEB), will cause a so called *in situ* generation of 1-hexene by oligomerisation of ethylene. This oligomerisation seems to be quite specific: during reactor runs with such a catalyst system, no increase in 1-butene or 1-octene was found in the flash gas analysis.
Besides the generation of 1-hexene, TEB will cause other effects, which may be advantageous: a modification of the molecular weight distribution, an increase or decrease of the melt index, and a change in catalyst activity. Those effects are different for each CrOx catalyst, and should be investigated on lab scale before full scale application.
The rate of generation of 1-hexene is higher when the CrOx catalyst has been reduced by CO at the end of its activation, and before its injection into the reactor. This may bring an important saving of the more expensive comonomer, but the molecular weight distribution of the polymer, reactor conditions unchanged, will be much more narrow, so the quality of the product should be completely revised.

To control the reactor with this CrOx/TEB catalyst system, the rate of 1-hexene generation and the 1-hexene conversion should be known.
The mass balance (9.2), rewritten for 1-hexene, should now take into account the *rate of in-situ 1-hexene generation*, R_6, and the *conversion of the in-situ generated 1-hexene*, ξ_{61}. There is still a feed flow F_6 of 1-hexene to the reactor, because, even if no fresh 1-hexene is fed, the recycle stream from the recovery section will contain some 1-hexene, as an absolute separation of 1-hexene from the recycled diluent stream is not practical.
The mass balance thus reads,

$$F_6(1-\xi_6) + R_6(1-\xi_{61}) = D_\text{R}\, y_6 \frac{84.16}{(\text{MW})_\text{fg}}\,.$$

When divided by the mass balance (9.3) for the isobutane diluent,

$$F_\text{IC4} = D_\text{R}\, y_\text{IC4} \frac{58.12}{(\text{MW})_\text{fg}}\,,$$

for a better accuracy, the following expression is obtained,

$$(1-\xi_6) + \frac{R_6}{F_6}(1-\xi_{61}) = \frac{F_\text{IC4}}{F_6}\frac{y_6}{y_\text{IC4}}\frac{84.16}{58.12}\,.$$

The right hand member of the expression is obtained from the slope of a y_6/y_{IC4} versus F_{IC4}/F_6 plot, as explained in the previous sections. Only, the much steeper slope will give the false impression of a low 1-hexene conversion, since the flash gas is containing more 1-hexene from the in-situ generation.
There are three unknown variables ξ_6, ξ_{61} and R_6, which cannot be solved from one equation.
Since it is not possible to discriminate between 1-hexene from the feeds and 1-hexene generated in the reactor, a *global 1-hexene conversion*, ξ_6^*, is defined as the conversion of all 1-hexene, either fed to the reactor or generated in situ. The mass balance for 1-hexene then looks like,

$$(F_6 + R_6)(1 - \xi_6^*) = D_R\, y_6 \frac{84.16}{(MW)_{fg}} \,,$$

or, after dividing by the mass balance for the isobutane diluent,

$$\left(1 + \frac{R_6}{F_6}\right)(1 - \xi_6^*) = \frac{F_{IC4}}{F_6}\frac{y_6}{y_{IC4}}\frac{84.16}{58.12} \,. \tag{9.13}$$

To bring extra information, it is assumed that the correlation between density and incorporated 1-hexene (9.12) for the CrOx product is known, and also valid when part of the 1-hexene is generated in-situ by TEB. Since the product density is known when treating historical data, or as a target value for projected products, the corresponding fraction of incorporated 1-hexene, x_6 is known by correlation (9.12). x_6 is related to the conversions accorded to the following adapted version of equation (9.11),

$$x_6 = \frac{(F_6 + R_6)\xi_6^*}{P} \,. \tag{9.14}$$

Eliminating ξ_6^* from (9.13) by use of (9.14) yields,

$$\frac{R_6}{F_6} = \frac{P x_6}{F_6} - 1 + \frac{F_{IC4}}{F_6}\frac{y_6}{y_{IC4}}\frac{84.16}{58.12} \,. \tag{9.15}$$

Eliminating R_6 from (9.13) by use of (9.15) yields,

$$\xi_6^* = 1 - \left(\frac{F_{IC4}}{F_6}\frac{y_6}{y_{IC4}}\frac{84.16}{58.12}\right)\left(\frac{P x_6}{F_6} + \frac{F_{IC4}}{F_6}\frac{y_6}{y_{IC4}}\frac{84.16}{58.12}\right)^{-1} \,. \tag{9.16}$$

So both ξ_6^* and R_6 can be estimated. A correlation for R_6 is not available here, by lack of data from sufficiently varying reactor conditions.
For ξ_6^*, a correlation of the type,

$$\xi_6^* = \beta_0 + \beta_1 y_2 + \beta_2 T \,,$$

is found, explaining some 63% of the total variance. The coefficient β_1 for the ethylene concentration is negative, and β_2, for the temperature, positive, as expected.
For small extrapolations from existing reactor conditions, and when little information is available, it has been assumed, with some success, that ξ_6^* is constant and that R_6 is proportional to the ethylene feed F_2.

9.7 Optimisation of Reactor Conditions Based on Conversions

Especially with Ziegler–Natta catalysts, due to their low 1-hexene conversion, the production of HDPE's with lower densities is problematic due to the high 1-hexene concentrations involved, and the vicinity of the fouling curve. By a proper optimisation of the reactor conditions, a drastic decrease of the 1-hexene concentration can be obtained, which renders the production of a grade of lower density possible at a respectful distance from the fouling curve.

The goal of the optimisation is to realise a maximal fraction of incorporated 1-hexene, i.e. a maximisation of x_6, at a 1-hexene concentration on the safe side of the fouling curve.

The incorporated 1-hexene is equal to, equation (9.11),

$$x_6 = \frac{F_6 \xi_6}{P}\,.$$

Since the mass fraction of the hydrogen can be neglected,

$$x_6 = \frac{F_6 \xi_6}{F_2 \xi_2 + F_6 \xi_6}\,,$$

whence follows the 1-hexene to ethylene ratio in the reactor feeds,

$$\frac{F_6}{F_2} = \frac{\xi_2}{\xi_6} \frac{x_6}{1 - x_6}\,.$$

Combined with the relation between the 1-hexene to ethylene ratio in the reactor feeds and in the flash-gas, equation (9.8),

$$\frac{F_6}{F_2} \frac{1 - \xi_6}{1 - \xi_2} = \frac{y_6}{y_2} \frac{84.16}{28.05}\,,$$

an expression for the 1-hexene concentration in the flash-gas is obtained,

$$y_6 = y_2 \frac{\xi_2}{\xi_6} \frac{1 - \xi_6}{1 - \xi_2} \frac{x_6}{1 - x_6} \frac{28.05}{84.16}\,.$$

Hence an expression for the fraction of incorporated 1-hexene, in terms of the other variables,

$$\frac{x_6}{1 - x_6} = \left(y_6 \frac{84.16}{28.05} \right) \left(\frac{1}{y_2} \frac{1 - \xi_2}{\xi_2} \frac{\xi_6}{1 - \xi_6} \right). \tag{9.17}$$

The left member of equation (9.17) is increasing monotonously with x_6, so the goal of the optimisation can be rephrased into the maximisation of the right hand member of the equation.

It is obvious that the concentration of 1-hexene should be maximal, so it will be taken equal to the maximum concentration of 1-hexene that is considered as a safe value in view of the fouling curve at the targeted density of the polymer.

The right group in the right member of the equation should thus be maximised in function of the other variables y_2, y_{cocat} and T.
For a Ziegler–Natta catalyst, the optimisation is based on the empirical correlations obtained for that catalyst, which own the same format,

$$\xi_2 = \beta_{2,0} + \beta_{2,1}y_2 + \beta_{2,2}y_{cocat} + \beta_{2,3}T\,,$$
$$\xi_6 = \beta_{6,0} + \beta_{6,1}y_2 + \beta_{6,2}y_{cocat} + \beta_{6,3}T\,.$$

The influence of the cocatalyst and the temperature are for both conversions positive and rather small. If p_2 is a small relative increase (p_2 being a little larger than 1) of ξ_2, and p_6 is a small relative increase of ξ_6, the relative increase of the group at constant ethylene concentration is,

$$\left(\frac{1-p_2\xi_2}{p_2\xi_2}\frac{p_6\xi_6}{1-p_6\xi_6}\right)\left(\frac{1-\xi_2}{\xi_2}\frac{\xi_6}{1-\xi_6}\right)^{-1} \rightarrow \frac{p_6}{p_2}\,,$$

since $(1-p_2\xi_2) \rightarrow (1-\xi_2)$, and $(1-p_6\xi_6) \rightarrow (1-\xi_6)$. For the particular catalyst that was studied in this respect, the increase of ξ_6 with y_{cocat} and T is relatively stronger than ξ_2, or $p_6/p_2 > 1$ for all ranges of ξ_2 and ξ_6, so it is indicated that both the cocatalyst concentration and the temperature should be maximised. The temperature is of course limited by the fouling curve.
To optimise the ethylene concentration, a similar reasoning could be adopted, whence it may be expected that a lower ethylene concentration will yield a larger value of the group. Still, this result is doubtful, as the effect of y_2 on the conversions is stronger, and the conversions cannot be regarded as independent of y_2. The most direct way is here the best approach, i.e. simply to calculate the group

$$\left(\frac{1}{y_2}\frac{1-\xi_2}{\xi_2}\frac{\xi_6}{1-\xi_6}\right)$$

at the already optimised conditions y_{cocat} and T from the correlations, plot it versus y_2, and see where the maximum is reached. For the same particular catalyst, the group shows a monotonous increase for decreasing ethylene concentrations, so it is advised to operate at the lowest ethylene concentration.

The results of this kind of optimisation are spectacular. One example is that the production of a Ziegler–Natta reactor powder with a very low density was made feasible by the application of this method. The ethylene concentration was reduced to a minimum. The 1-hexene concentration was 24% lower than required at standard conditions, and located above the fouling curve. The required total 1-hexene feed was reduced by some 20%, thus solving a problem of 1-hexene feed capacity, that would have arisen without the optimisation.

Notations

The reactants and chemical substances are indicated by a subscript:

2	ethylene
6	1-hexene
H	hydrogen
i	one of the reactants ethylene, 1-hexene or hydrogen
IC4	isobutane
cat	catalyst
cocat	cocatalyst

DENS	density of the polymer measured in the lab at room temperature, (kg, m^{-3}).
D_R	flow of recycled diluent (reactor liquid, or flash gas) leaving the reactor, (kg, s^{-1}).
F_2	ethylene feed to the reactor, (kg, s^{-1}).
F_6	1-hexene feed to the reactor, (kg, s^{-1}).
F_H	hydrogen feed to the reactor, (kg, s^{-1}).
F_{IC4}	isobutane diluent feed to the reactor, (kg, s^{-1}).
MFI	melt index of the polymer, (dg, min^{-1}).
$(MW)_{fg}$	average molecular weight of the flash gas, leaving the reactor, ($kg, kmole^{-1}$).
P	polymer production rate in a reactor: total quantity of polymer produced in the reactor per unit of time, (*kg* polymer, s^{-1}).
R_6	rate of in-situ 1-hexene generation by TEB with a CrOx catalyst, (kg, s^{-1}).
T	reactor temperature, (°C).
T_{act}	chromium oxide catalyst activation temperature, (°C).
x_6	incorporated 1-hexene in the polymer, (*kg* 1-hexene/*kg* polymer).
x_H	incorporated hydrogen in the polymer, (*kg* hydrogen/*kg* polymer).
y_2	ethylene concentration in the flash gas, (*mole* ethylene/*mole* flash gas).
y_6	hexene concentration in the flash gas, (*mole* 1-hexene/*mole* flash gas).
y_{cocat}	cocatalyst concentration in the reactor liquid, (*kg* cocatalyst/*kg* reactor liquid).
y_H	hydrogen concentration in the flash gas, (*mole* hydrogen/*mole* flash gas).
y_{IC4}	isobutane concentration in the flash gas, (*mole* isobutane/*mole* flash gas).
ξ_2	ethylene conversion in the reactor: the fraction of all ethylene that is fed to the reactor, which is consumed in the polymerisation reaction, (–).

ξ_6	1-hexene conversion in the reactor: the fraction of all 1-hexene that is fed to the reactor, which is consumed in the polymerisation reaction, (–).
ξ_{61}	conversion of the in-situ generated 1-hexene: the fraction of all 1-hexene that is generated in the reactor by the use of a cocatalyst, which is consumed in the polymerisation reaction, (–).
ξ_6^*	over all 1-hexene conversion in the reactor: the fraction of all 1-hexene that is fed to the reactor or generated in the reactor by the use of a cocatalyst, which is consumed in the polymerisation reaction, (–).
ξ_H	hydrogen conversion in the reactor: the fraction of all hydrogen that is fed to the reactor, which is consumed in the polymerisation reaction, (–).
ξ_{IC4}	isobutane diluent conversion in the reactor, which is zero, (–).

10 The Ethylene Concentration Profile in the Loop Reactor

In the next Chapter 11, a simple and practical model will be built, intended to answer questions as to what reactor conditions and settings are required to produce a certain product. One single value of each variable will be assigned to the reactor system, without considering any distribution of these variables inside the loop. But before doing so, and to gain more confidence in the model, some insight is needed into what is happening in the loop reactor when a product is produced at a known set of reactor conditions, and more specifically, what the ethylene concentration profile looks like.

This will be the only chapter where is looked inside the loop, and where a limited number of variables will vary along its length.

It is assumed that certain parameters are one-dimensionally distributed, i.e. vary along the length of the reactor tube, but at the same time are uniform throughout any cross-sectional area of the reactor. This means that perfect mixing is supposed to occur in radial directions over the cross-sectional area.

A second, and even stronger assumption is made by stating that there is no mixing at all in the axial direction, along the length of the reactor tube. This is acceptable when the very high axial velocity and fully developed turbulent flow conditions of the slurry are considered.

Further assumptions are, steady state operation, the 1/1 hypothesis, and perfect mixing for the catalyst.

The discussion is also limited to the production of a homopolymer, so no comonomer is taking part in the reaction.

It should be stressed that this chapter is treating one of the few arguments in this book, that have not been confirmed experimentally. Indeed, concentration profiles are not measured on commercial loop reactors. Still it is tried to make the assumptions comply with the knowledge gained in the previous chapters, and with what is generally accepted in the literature.

10.1 Reactor Geometry

The reactor geometry used in this discussion is extremely simplified: the closed loop owns a *length* l, and a *position along that length* is denoted as z.

There is only one feed point where all feeds enter the reactor. It owns no length, and is located at $z = 0$. Because the loop is closed, the point $z = l$ will coincide with $z = 0$.

There is only one take-off point where part of the slurry is leaving the reactor as the product stream. It has no length either, even if in reality a number of settling legs are spread over a limited length of the reactor tube. Its position is $z = z_1$.

The stream of slurry through the loop will be diminished when it has passed the take-

off point, as it has lost the product stream. This reduced stream starts at the position $z = z_2$. This position coincides with z_1, or, $z_2 = z_1^+$. The loop is thus divided in two parts, *part 1* from $z = 0$ till $z = z_1$, and *part 2* from $z = z_2$ till $z = l$, where the slurry flow is somewhat smaller.
The *cross sectional area* of the loop tube is Ω, and thus the *volume* of the reactor is $V_R = \Omega l$.

10.2 Slurry Flow in the Reactor

It is important to note that, for the sake of simple calculations, all quantities and concentrations are expressed in units of weight.

The slurry circulated through the loop is the sum of two fluxes, a flux of liquid and a flux of solids.
The *liquid flux*, $L(z)$, (kg, m^{-2}, s^{-1}), is containing the liquid components isobutane (or another diluent), and ethylene. No comonomer is present, and hydrogen is discarded. The concentration of ethylene therein, y_2, is expressed as a weight fraction.
The *solids flux*, $S(z)$, (kg, m^{-2}, s^{-1}), is composed of the polymer and the catalyst.

In each part of the reactor, part 1 and part 2, the total mass flux through the loop is constant,

$$S + L = c\,. \tag{10.1}$$

Boundary conditions for S and L at the feed- and take-off points are,

$$\begin{aligned} L(0) &= L(l) + \Sigma F_i/\Omega\,; \\ S(0) &= S(l)\,; \\ L(z_2) &= L(z_1) - D_R/\Omega\,; \\ S(z_2) &= S(z_1) - P/\Omega\,; \end{aligned}$$

where, ΣF_i is the sum of all feeds to the reactor, D_R is the liquid stream recycled to the reactor, and P is the production rate. The catalyst feed stream is neglected.
The sum of the four boundary conditions yields the over-all mass balance for the reactor, $\Sigma F_i = D_R + P$.

The *velocity of the slurry*, v, is,

$$v = \frac{S}{\rho_p} + \frac{L}{\rho_f}\,. \tag{10.2}$$

Here it is not supposed, like in Chapter 3, that the densities of the polymer and the liquid, ρ_p and ρ_f, are constant. If so, the liquid would be incompressible, which it is not, and the effects of pressure and liquid composition on ρ_f would be disregarded. Instead, *it is assumed that the slurry velocity v is constant along the whole loop*. This is not only convenient in calculations, but also in accordance with the experience that the circulation period of the slurry is constant, and that the liquid is quite compressible.

In process computer calculations, it is usually assumed that also the *slurry density*, ρ_S, is constant along the loop, although it is only measured at one or two points at the loop.
The mass flux through a slab Ωdz of the reactor is,

$$\frac{1}{\Omega}\frac{dm}{dt} = \frac{\rho_S \Omega dz}{\Omega dt} = \rho_S v\,,$$

hence,

$$S + L = \rho_S v\,.$$

The simultaneous assumptions of a constant slurry velocity and a constant slurry density lead to a contradiction at the take-off point where,

$$(S + L)_{z=z_2} = (S + L)_{z=z_1} - \frac{P + D_R}{\Omega}\,.$$

The contradiction could be solved by accepting the value of ρ_S in the part of the reactor where it is measured, and allowing a slightly different value for ρ_S in the other part, $\rho_S \pm (P + D_R)/\Omega v$. Besides, the difference $(P + D_R)/\Omega$ is, in commercial operation, only of the order of 1% of the total mass flux $\rho_S v$. So assuming constant slurry velocity and density at the same time is quite acceptable.

10.3 Catalyst Activity

In this discussion, the catalyst activity is assumed to be variable, not only due to the lifetime of the catalyst particle, but also due to a varying ethylene concentration $y_2(z)$ in the loop.
This relation between the catalyst activity and the ethylene concentration is based on the literature, where it is generally accepted that the catalyst activity is proportional to the ethylene concentration, see Section 5.2 in Chapter 5.

An expression for the catalyst activity should be reconciled with the empirical findings from Chapter 8, where one catalyst activity profile was found for the whole reactor loop, equation (8.4),

$$\overline{A}(\theta) = be^{-a\theta}\,,$$

at a certain set of reactor conditions, among which the ethylene concentration $y_2(z = z_1)$, measured at the take-off point $z = z_1$, is an important condition[1].

---

1 In practical plant operation, the ethylene concentration measured in the flash gas analysis is accepted as the variable representing the ethylene concentration in the whole reactor. The flash gas stream is identical to the liquid in the product stream, taken from the reactor at the reactor take-off point, usually via the settling legs.

This suggests the introduction of an average ethylene concentration $\overline{y}_2$, where,

$$\overline{y}_2 = \frac{1}{l}\int_{z=0}^{l} y_2(z)\mathrm{d}z\,. \tag{10.3}$$

The choice of a linear average is based on the first order rate of the polymerisation reaction.

A modified expression for the catalyst activity, based on equation (8.4), is then postulated as,

$$\overline{\mathcal{A}}(\theta) = \frac{b}{\overline{y}_2} y_2(z)\mathrm{e}^{-a\theta}\,. \tag{10.4}$$

Catalyst activity is thus dependent on the position z in the loop, but still in accordance with equation (8.4) since, if the ethylene concentration were constant in the whole loop, $y_2(z) = y_2 = \overline{y}_2$, its expression would be reduced to (8.4).

The catalyst reactor activity, $\mathbb{A}$, has been defined in Section 5.3 for the whole reactor as the rate of polymer produced by one *kg* of catalyst present in that reactor per unit time. This definition is now adapted for the catalyst present in a small slab $\Omega\mathrm{d}z$ of the reactor, because the reactor activity is now also dependent on z.

Adapting equation (7.2),

$$\mathbb{A}(z) = \int_{\theta=0}^{\infty} E_2(\theta, Rx)\overline{\mathcal{A}}(\theta, Rx, z)\mathrm{d}\theta = \int_{\theta=0}^{\infty} \frac{1}{\tau}\exp\left(-\frac{\theta}{\tau}\right)\frac{b}{\overline{y}_2}y_2(z)\mathrm{e}^{-a\theta}\,,$$

or,

$$\mathbb{A}(z) = \frac{y_2(z)}{\overline{y}_2}\frac{b}{1+a\tau}\,. \tag{10.5}$$

The average reactor activity over the whole length of the loop is still in accordance with the expression 8.5 obtained earlier for the catalyst reactor activity,

$$\mathbb{A} = \frac{1}{l}\int_{z=0}^{l} \mathbb{A}(z)\mathrm{d}z = \frac{1}{\overline{y}_2}\frac{b}{1+a\tau}\frac{1}{l}\int_{z=0}^{l} y_2(z)\mathrm{d}z = \frac{b}{1+a\tau}\,.$$

It is allowed to average the catalyst activity over the length of the loop, because, due to the assumption of perfect mixing of the catalyst, the catalyst concentration is the same along the whole length of the loop.

It is also necessary to verify whether the definition (10.4) of catalyst activity is in line with the expressions for catalyst productivity that have been obtained earlier.

The catalyst particle productivity has been defined in Section 6.3 as the quantity of polymer amassed by the catalyst particle on its own surface during the course of its lifetime in the reactor, equation (6.1),

$$\mathcal{P} = \int_{t=0}^{\theta} \overline{\mathcal{A}}(t, Rx)\mathrm{d}t = \int_{t=0}^{\theta} \overline{\mathcal{A}}(z(t), t, Rx)\mathrm{d}t\,.$$

During its lifetime t, the particle is circulated through the loop at a period that is very small compared to the average catalyst residence time. The position of the particle $z(t)$ is periodic and changing so fast that the catalyst activity $\overline{\mathcal{A}}(z(t), t, Rx)\mathrm{d}t$ can be replaced by the average catalyst activity, equation (10.4),

$$\overline{\mathcal{A}}(\theta) = \frac{b}{\overline{y}_2}\overline{y}_2(z)\mathrm{e}^{-a\theta} = b\mathrm{e}^{-a\theta}$$

The catalyst particle productivity and the catalyst reactor productivity are thus equal to the expressions found earlier in Section 8.3.

10.4 Rate of Reaction along the Loop

The rate of polymerisation will be variable along the loop. It is equal to the product of the quantity of catalyst present and the catalyst activity.

The total mass of catalyst present in the reactor is already known from the total production rate P and the average catalyst reactor activity $\mathbb{A}$, equations (7.10) and (8.5),

$$P = m_{\mathrm{cat}}\mathbb{A} \Rightarrow m_{\mathrm{cat}} = \frac{P(1 + a\tau)}{b} .$$

Due to the assumption of perfect mixing for the catalyst, the concentration of catalyst is constant along the loop. The *catalyst concentration in the reactor*, C_{cat}, is an interesting parameter, and is equal to,

$$C_{\mathrm{cat}} = \frac{m_{\mathrm{cat}}}{V_{\mathrm{R}}} = \frac{P(1 + a\tau)}{bV_{\mathrm{R}}} . \tag{10.6}$$

The mass of catalyst present in a slab $\Omega\mathrm{d}z$ of the reactor is thus,

$$\mathrm{d}m_{\mathrm{cat}} = \frac{P(1 + a\tau)}{bV_{\mathrm{R}}}\Omega\mathrm{d}z = \frac{P(1 + a\tau)}{bl}\mathrm{d}z . \tag{10.7}$$

The rate of polymerisation in a slab $\Omega\mathrm{d}z$ is then, equation (10.5),

$$r(z) = \mathrm{d}m_{\mathrm{cat}}\mathbb{A}(z) = \frac{P(1 + a\tau)}{bl}\mathrm{d}z\frac{y_2(z)}{\overline{y}_2}\frac{b}{1 + a\tau} ,$$

or,

$$r(z) = P\frac{y_2(z)}{\overline{y}_2}\frac{\mathrm{d}z}{l} . \tag{10.8}$$

This simple expression reflects the assumptions that the catalyst is evenly spread in the loop, and that its activity is proportional to the ethylene concentration.

The detailed expression for catalyst activity, with the constants a and b, is wrapped up in the production rate P, which makes that this expression for the rate of reaction is more generally valid, as long as there is perfect mixing for the catalyst, and that the catalyst activity is proportional to the ethylene concentration.

10.5 Ethylene Concentration Profile in the Loop

A differential equation describing the ethylene concentration profile along the loop can now be written, based on an ethylene mass balance for a slab Ωdz of the reactor,

$$\Omega L y_2 + \frac{d}{dz}(\Omega L y_2)dz = \Omega L y_2 - r(z) .$$

Or, using equation (10.8),

$$y_2 \frac{dL}{dz} + L \frac{dy_2}{dz} + \frac{P}{V_R} \frac{y_2}{\overline{y}_2} = 0 .$$

In this equation, the liquid flux $L(z)$ can be expressed with reference to its value $L(0)$ at the feed point $z = 0$, considering that in each slab Ωdz, a quantity of ethylene, equal to $r(z)$, is, by polymerisation, transferred from the liquid phase to the solid phase. Thus,

$$dL(z) = -dS(z) = -\frac{r(z)}{\Omega} = -\frac{P}{V_R} \frac{y_2}{\overline{y}_2} dz, \tag{10.9}$$

and,

$$L(z) = L(0) - \frac{P}{V_R} \frac{1}{\overline{y}_2} \int_0^z y_2 dz .$$

After solving y_2 from the equation, $L(0)$ should be determined from the boundary conditions at the feed point, based on the information found above in Section 10.2. Upon substitution of L and dL, the equation reads,

$$y_2^2 - y_2 + \left(\int_0^z y_2 dz \right) \frac{dy_2}{dz} - \frac{L(0) V_R \overline{y}_2}{P} \frac{dy_2}{dz} = 0 .$$

The equation is made dimensionless by introducing a *dimensionless position*[2], s,

$$s = \frac{z}{l} .$$

Hence,

$$y_2^2 - y_2 - \left(K - \int_0^s y_2 ds \right) \frac{dy_2}{ds} = 0, \tag{10.10}$$

where K is a dimensionless number,

$$K = \frac{L(0) \Omega \overline{y}_2}{P} . \tag{10.11}$$

2 Note that y_2, expressed in weight fraction, is already dimensionless.

The solution of this equation requires numerical methods that are outside the scope of this book[3].

Still, an important conclusion is that the ethylene concentration profile along the loop is governed by only one dimensionless number, K, Equation 10.11, which may be called *the ethylene recycle ratio*, as it is equal to the ratio of the rate of ethylene circulated through the loop, versus the rate of ethylene consumed in the production. The larger this number, the more flat the concentration profile will be. High values of this number applied in commercial loop reactors make them approach continuous stirred tank behaviour, where the ethylene concentration would be uniform throughout the reactor volume.

If an analytical solution of equation (10.10) is not straightforward, it is satisfactory to consider an approximate solution for the ethylene concentration profile, considering that the value of the liquid flux L along the loop is nearly constant.

Starting from $z = 0$, L is constantly decreasing by the loss of ethylene to the reaction, and the total variation of L, along the whole loop, including parts 1 and 2, is, equation (10.9),

$$\Delta L = \int_{z=0}^{l} \mathrm{d}L = -\int_{z=0}^{l} \frac{P}{V_R}\frac{y_2}{\bar{y}_2}\mathrm{d}z = -\frac{P}{\Omega}\,.$$

Now, this value P/Ω is, in commercial operation, only of the order of 1% of the liquid flux, and L may be considered as approximately constant.

Its value may be estimated as,

$$L \simeq \rho_S v(1 - C_W), \tag{10.12}$$

where C_W is *the solids concentration in the reactor, expressed as a weight fraction.* A minor correction $\pm D_R/\Omega$ may perhaps be made for part 1 or part 2 of the reactor, as explained in Section 10.2.

The differential equation describing the ethylene concentration profile along the loop can then be rewritten as,

$$\Omega L y_2 + \Omega L \mathrm{d}y_2 = \Omega L y_2 - r(z),$$

or,

$$\mathrm{d}y_2 = -\frac{P}{\Omega L \bar{y}_2} y_2 \mathrm{d}s,$$

and easily solved as,

$$y_2(s) = y_2(s_1)\exp\left(\frac{s_1 - s}{K}\right), \tag{10.13}$$

3 A transformation of the variable y_2 into $y = \int_0^s y_2 \mathrm{d}s$ yields a homogeneous, non-linear, second order differential equation in y, where the variable s is not present, $y''(y - K) + (y')^2 - y' = 0$.

where K is the ethylene recycle ratio,

$$K = \frac{L\Omega\overline{y}_2}{P}, \tag{10.14}$$

and the boundary condition has been taken into account at the take-off point where the ethylene concentration $y_2(s_1) = y_2(z_1) = y_2(z_2)$, is measured in the flash gas.

The average ethylene concentration $\overline{y}_2$ follows then,

$$\overline{y}_2 = \int_{s=0}^{1} y_2(s)\mathrm{d}s = K\left[\exp\left(\frac{s_1}{K}\right) - \exp\left(\frac{s_1 - 1}{K}\right)\right] y_2(s_1)\,. \tag{10.15}$$

The solution for y_2 is found by solving $\overline{y}_2$ from (10.15), by iteration with a starting value $\overline{y}_2 = y_2(s_1)$, whereafter K and $y_2(s)$ are known.

Approximate ethylene concentration profiles have thus been calculated for different stable reactor conditions established when measuring the experimental catalyst productivity curve for CrOx catalyst nr. 2, as was described above, in Section 8.4, and shown in Figure 8.4.

The results are presented in Figure 10.1. For production rates ranging from 15.1 to 23.5 *tonnes per hour*, the ethylene concentration profiles are quite similar. The increases in production rate have been realized at approximately the same ethylene concentrations. Only at the highest production rate of 27.3 *tph*, did the operator apply a much higher ethylene concentration, which he believed was necessary to reach that rate. The data shows that the catalyst activity is still below expectations, in spite of his action. A closer look at this problem is taken in the next Section 10.6.

The ethylene concentration profiles are nearly linear and their slopes are approximately equal to,

$$\frac{\mathrm{d}y_2(s)}{\mathrm{d}s} \simeq \left(\frac{\mathrm{d}y_2(s)}{\mathrm{d}s}\right)_{s=s_1} = -\frac{y_2(z_1)}{K} = -y_2(z_1)\frac{P}{L\Omega\overline{y}_2}\,.$$

So they are always negative, as ethylene is consumed along the loop, and greater at higher production rates, as then, more ethylene is consumed. The great values of the liquid flux maintained in commercial operation, and, in one reactor, quite constant at all production rates, attenuate those slopes a great deal. The curves shown in Figure 10.1 show total drops in ethylene concentration over the whole loop from only 0.5 at the lowest to 1.0 *wt%* at the highest production rate.

It is useful to have an idea of what the ethylene concentration looks like along the loop. Extreme values of that concentration could expose the operation to a double risk, either in that the reaction be starved at low ethylene concentrations near the end of the loop ($z \to l$), or in that the bubble point of the liquid be reached near the feed point ($z \to 0$). The practice of maintaining a high liquid flux has, among others, the advantage that the ethylene concentration profiles are quite flat, and that the value of the ethylene concentration measured in the flash gas, at the take-off point, is representative for the concentration in the whole reactor. To avoid the risks just mentioned it is clearly recommended to keep this concentration within narrow limits.

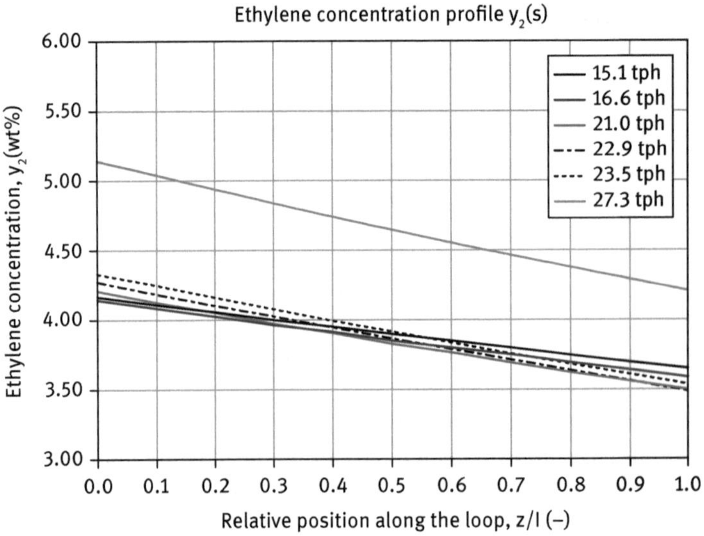

Fig. 10.1. Approximate ethylene concentration profiles at different reactor conditions established when measuring the experimental catalyst productivity curve for CrOx catalyst nr. 2 shown in Figure 8.4.

10.6 Increasing Production Rates in the Loop Reactor

If, in the previous section, it is explained that the ethylene concentration cannot be changed very much, it is interesting to see what actually does happen when the production rate in the reactor is increased.

An increase in solids concentration should extend the residence time of the catalyst, and thus its productivity, and so boost the production rate. But it can reasonably be expected that the solids concentration is already kept at a high level, where it can hardly be increased without risk, to obtain an economic, and also stable operation. So the solids concentration cannot be changed very much either.

The action taken by the operator when asked to increase the production rate, is to increase the catalyst feed rate F_{cat}, and, when the production rate P picks up, to increase the ethylene feed rate in order to maintain the ethylene concentration. Meanwhile, the solids concentration is kept constant by the proper manipulation of the diluent feed stream and the number of settling legs in service.

What is happening in the reactor can be read from the following table, which gives more information on the already mentioned stable reactor conditions for CrOx catalyst nr. 2, from Section 8.4, and Figure 8.4.

Production rate, P, *tph*	15.1	16.6	21.0	22.9	23.5	27.3
Ethylene Recycle Ratio, K	7.61	6.97	5.40	5.01	4.95	5.00
Catalyst Residence Time, τ, *h*	1.08	0.93	0.80	0.73	0.70	0.61
Cat. concentration, m_{cat}/V_R, gm^{-3}	37.2	44.5	54.4	64.0	75.4	89.4
Reactor Cat. Activity, $\mathbb{A}$, $kgh^{-1}kg^{-1}$	4,690	4,300	4,440	4,120	3,590	3,530
Cat. Productivity, $\mathbb{P}$, $kgkg^{-1}$	5,050	4,000	3,580	3,010	2,520	2,160

The catalyst residence time is decreased by the increased production rate P at a constant mass of solids m_{pol} according to equation (7.9),

$$\tau = \frac{m_{\text{pol}}}{P} .$$

The catalyst activity will not fall at shorter residence times, since the activity profile of the individual catalyst particles, $\overline{A} = be^{-a\theta}$, even tends to increase at lower residence times. When making this statement, the two cases at the highest production rates, 23.5 and 27.3 *tph*, are not taken into account.

The catalyst productivity is strongly reduced by the shorter residence times as there is less time to be productive.

The mass of catalyst in the reactor, or the catalyst concentration, is increased very much, in a ratio that is more than compensating the loss in productivity. The over-all result is an increase in production rate, but in a ratio that is less than proportional to the increase of catalyst concentration.

It seems obvious that the operator's action, an increase in catalyst feed, has led to an increase in catalyst concentration. But that is not straightforward in a reactor where the catalyst is contained in the mass of polymer, which is kept at a constant level, and where the residence time is affected by an increased production rate. According to equation (7.9),

$$F_{\text{cat}}\tau = m_{\text{cat}},$$

a higher catalyst feed rate will only increase the catalyst concentration if the residence time is not reduced too much, or, assigning a subscript 1 and 2 to the original and the newly established conditions resp.,

$$\frac{\tau_1}{\tau_2} < \frac{(F_{\text{cat}})_2}{(F_{\text{cat}})_1} .$$

Or, at constant m_{pol}, using equation (7.9),

$$\frac{P_2}{P_1} < \frac{(F_{\text{cat}})_2}{(F_{\text{cat}})_1},$$

which means that an increased catalyst feed rate will increase the catalyst concentration if the production rate is increased in a rate that is less than proportional. If not, the increased production rate will lower the catalyst residence time so much that the extra catalyst feed does not bring any more catalyst mass that stays in the reactor.
Making use of equation (7.11),

$$\mathbb{P} = \frac{P}{F_{\text{cat}}},$$

the condition to increase the catalyst mass is simply,

$$(\mathbb{P})_2 < (\mathbb{P})_1 ,$$

or, that the catalyst productivity in the new reactor condition is lower than in the original condition.
This is always the case, since the catalyst productivity curve always falls with lower residence times.

So, to increase the production rate in the loop reactor, a higher concentration of catalyst is established in the reactor, which, without a loss in catalyst activity, will yield a higher production rate. A loss in catalyst productivity should thereby be taken into account, due to the reduced catalyst residence time.

However, as seen from the example from CrOx catalyst nr. 2, Section 8.4, Figure 8.4, the catalyst productivity at the highest production rates, 23.5 and 27.3 *tph*, is far below what is expected from the catalyst productivity curve.
A similar loss of productivity is seen with CrOx catalyst nr. 1, Section 8.4, Figure 8.5. Since the major change at higher production rates in the reactor, is a strong increase of the catalyst concentration, whilst the ethylene concentration remains unchanged, it is highly probable that the loss in catalyst productivity is due to a loss in catalyst activity because the diffusion of ethylene through the liquid to the catalyst surface has become insufficient to feed the expected reaction rate.
When running very high production rates at high catalyst concentrations, an extra economic loss due to inefficient use of the catalyst must be taken into account.

Notations

a — constant in the expressions for the catalyst activity profile and catalyst productivity, (s^{-1}).

$\mathbb{A}$ — catalyst reactor activity: quantity of polymer produced by all catalyst present in the reactor, or in a certain part of the reactor, per unit time, per *kg* catalyst, (*kg* polymer/*kg* catalyst, *s*).

$\overline{\mathcal{A}}(t, Rx)$ — average catalyst activity over all catalyst particle sizes: quantity of polymer produced per *kg* of catalyst, per unit time, at a certain lifetime $t = \theta$ in the reactor, (*kg* polymer/*kg* catalyst, *s*).

b	constant in the expressions for the catalyst activity profile and catalyst productivity, (kg, kg^{-1}, s^{-1}).
C_{cat}	catalyst concentration in the reactor per unit volume, (kg, m^{-3}).
C_W	solids concentration in the reactor, expressed as a weight fraction, (–).
D_R	total flow of recycled reactor liquid (or flash gas) leaving the reactor, (kg, s^{-1}).
$E_2(\theta)$	catalyst residence time distribution, (s^{-1}).
F_{cat}	catalyst feed to the reactor, (kg, s^{-1}).
K	dimensionless number governing the ethylene concentration profile along the length of the reactor tube, (–).
l	total length of the reactor tube, (m).
$L(z)$	flux of liquid circulated through the loop, (kg, m^{-2}, s^{-1}).
m	mass, (kg).
m_{cat}	mass of catalyst present in the reactor, (kg).
m_{pol}	mass of polymer present in the reactor, (kg).
P	polymer production rate: total quantity of polymer produced in the reactor per unit of time, (kg polymer, s^{-1}).
$\mathbb{P}$	catalyst reactor productivity: total quantity of polymer produced by one kg of catalyst that has left the reactor, (kg polymer/kg catalyst).
$\mathcal{P}$	catalyst particle productivity, (kg polymer/kg catalyst).
$r(z)$	rate of polymerisation in a slab Ωdz of the reactor, (kg polymer, s^{-1}).
Rx	symbol representing all of the reactor conditions, including temperature, concentrations, flows, and densities.
s	dimensionless position along the length of the reactor tube, (–).
$S(z)$	flux of solids circulated through the loop, (kg, m^{-2}, s^{-1}).
v	velocity of the slurry, (m, s^{-1}).
V_R	reactor volume, (m^3).
y_2	ethylene concentration, (kg ethylene/kg reactor liquid).
$\overline{y}_2$	average ethylene concentration in the reactor, (kg ethylene/kg reactor liquid).
y_{cat}	catalyst concentration, (kg catalyst/kg reactor solids).
z	position along the length of the reactor tube, (m).
θ	catalyst residence time: the total time that a catalyst particle has spent in the reactor at the moment when it leaves the reactor, (s).
ρ_f	density of the liquid in the reactor, (kg, m^{-3}).
ρ_p	density of the polymer in the reactor, (kg, m^{-3}).
ρ_s	density of the slurry in the reactor, (kg, m^{-3}).
ΣF_i	sum of all feeds to the reactor, (kg, s^{-1}).
τ	average catalyst residence time, (s).
Ω	internal cross sectional area of the loop reactor tube, (m^2).

11 A Simple Model for the Slurry Loop Reactor

Enough information has been collected in the previous chapters to construct a simple model which describes the functioning of the loop reactor. Simple, because all variables are supposed to own only one value at a certain point in time. There is no spatial distribution considered for the variables over the volume of the reactor, just like in the classic model of the continuous stirred tank reactor, where all variables, like temperature and concentrations, are supposed to be constant throughout the volume of the tank.

The model can be put into practice with the help of a commercial spread-sheet program, and however simple it may be, it is efficient as an off-line model to plan or to steer full scale reactor runs. It cannot be expected to give always the highest precision, and extrapolations are certainly risky, because there are always unknown factors, like catalyst poisons, that influence the performance of the catalyst. Still, to cope with uncertainties, the model itself can be used to construct algorithms that yield corrections based on stable live reactor conditions.

The following description and construction of the model offer an example of how, in the author's opinion, models should be constructed.

First comes a careful listing of all variables describing the system. Then the list of variables is split up in two groups, independent variables, which can be chosen or influenced by the operator, and dependent variables, which follow from the values of the independent variables. It is often surprising how very few variables can really be considered as independent variables.

Then a number of mathematical relations should be established that allow to calculate all dependent variables from the independent variables. Simplifying assumptions will often facilitate this step, and there is nothing wrong with that, if those assumptions are explicitly mentioned. Assumptions are often made tacitly, or even unknowingly. When they are expressed directly, the user of the model is fully aware of them, as well as of the limitations inherent to the model. They allow to have a full knowledge of the weaknesses and flaws in the model, and to avoid much unnecessary criticism.

The last step is to solve the mathematical relations, and make the solutions available in a user-friendly spreadsheet or program.

The following basic assumptions are made in this model.

1. Steady state operation.
2. The 1/1 hypothesis is valid: each catalyst particle forming one polymer particle. See Chapter 6.
3. Perfect mixing for the catalyst: even if the polymer particles are not perfectly mixed, the catalyst particles enclosed in the polymer particles are so. See Section 7.2.

11.1 Elements to Build the Model with

11.1.1 Overview of the Variables

The following elements are used to build the model. They can be regarded as a number of choices that can be made, or a number of degrees of freedom which can be fixed, and which will determine the functioning of the reactor.

1. The reactor geometry. In this model, the volume V_R of the reactor is of use, the number N of settling legs in service, and the diameter D of the settling legs. The length of the reactor is not involved, since no distributions of variables along the reactor length are considered.
2. The type of diluent used (usually isobutane or propane).
3. The reactor pressure p. The reactor pressure is usually set at a certain value, and never changed. Together with the choice of the diluent, it is so constant a value that it is not explicitly mentioned in the reactor conditions.
4. The reactor conditions, represented as Rx. They should be listed in detail as the reactor temperature T; the concentrations in the diluent of ethylene y_2, hexene y_6, and hydrogen y_H; the catalyst and cocatalyst concentrations y_{cat} and y_{cocat}; the solids concentration C_0; the polymer density ρ_p and the liquid density ρ_f. Concentrations of other components are disregarded, even if they can be important (e.g. poisons are supposed not to be present).
   For chromium oxide (CrOx) catalysts, the catalyst activation temperature T_{act} should be added to the list of reactor conditions, even if the activation is effected outside the reactor and before the polymerisation.
5. The type of catalyst and cocatalyst. The selection of the catalyst and cocatalyst type will fix the catalyst activity profile $\overline{A}(\theta, Rx)$ at a certain set of reactor conditions Rx.
6. The polymer properties. The catalyst will, at the reactor conditions Rx, determine the polymer quality.
   Polymer properties are the polymer melt index MFI, the density DENS, as measured in the laboratory, the polymer density in the reactor, ρ_p, the molecular weight distribution, the incorporated ethylene, 1-hexene and hydrogen in the polymer, x_2, x_6, and x_H, the distribution of the comonomer over the chains of the polymer. The catalyst system will, at the reactor conditions Rx, also determine the powder morphology, which will determine the settling velocity $v_{sett}(Rx)$ of the powder particles in the settling legs.
7. The reactor feeds. The diluent feed F_{dil}, the ethylene feed F_2, the hexene (or any other comonomer) feed F_6, the hydrogen feed F_H, the catalyst feed F_{cat} and cocatalyst feed F_{cocat}.
8. The mass of solids in the reactor, m_{pol}, and the mass of catalyst in the reactor, m_{cat}. The catalyst residence time distribution is known based on the assumptions

of 1/1 hypothesis and perfect catalyst mixing. Thus it is reduced to one single variable, the average catalyst residence time τ. See Section 7.6.

9. The catalyst reactor activity $\mathbb{A}$, the catalyst reactor productivity $\mathbb{P}$, the production rate P.

11.1.2 Independent Variables

Solving a set of simultaneous equations is one of the basic skills in algebra. When the number of independent equations is equal to the number of variables, or "unknowns", there is a unique solution. When it is smaller, the system is underdetermined. When there are more equations than variables, the system is overdetermined, and there is no solution.

When solving real life problems, and especially technical problems, it is good to have this scheme in mind. Only, there is no teacher in the classroom telling which variables are known, and which are the unknowns...

The fact that the reactor is running, and producing only one polymer of a certain melt index and density at the time, means that the system of equations, formed by the existing relations between the variables, must be determined, since its unique solution is visible in the plant. Building a model then, is nothing else than writing down the relations governing the system as formal equations between the variables of the system.

The variables have already been listed. Their number is greater than the number of equations that is available. The next step is to assign to all listed variables a status of independent, or dependent variable. By *independent variables* are meant those to which a certain value is given, thus reducing the number of variables in the model. *Dependent variables* are those that are left, as "unknowns", and of which the value must be calculated from the available equations. Their number must be equal to the number of those equations.

Sometimes there is an objective reason to call a variable dependent or independent. The reactor volume e.g. is an independent variable. Its value has been fixed before the plant was constructed.

But the status of the variables is most often based on a mode of reactor operation, or rather, operating philosophy. There must be no confusion with process control loops, where the status of manipulated variable, resp. controlled variable, does not necessarily coincide with that of independent or dependent variable in the model. If e.g. the operator decides to run the reactor at a certain temperature, then that temperature is an independent variable to the model, even if it is controlled by a temperature control loop. This holds also for variables like solids concentration or ethylene concentration, which are controlled by the process control system, but can be selected as independent variables to the model by the mere decision on a setpoint value for those variables. The status of independent or dependent variable is not defined by a relation of cause and effect either. The melt index e.g. is here listed as an independent variable, because its target value has been selected before the reactor is run, as it is an important requirement to the product. It is the effect of several causes, like incorporated hydrogen, temperature, incorporated comonomer, and activation temperature (for a CrOx catalyst). However, to manage the system well, it is wise to control the melt index with the strongest agent, temperature for a CrOx catalyst, or hydrogen for a Ziegler–Natta catalyst, which is then a dependent variable, as its value is a function of the targeted melt index. The other variables that influence the melt index to a lesser degree, are kept at a fixed value as independent variables. Different causes to the same effect can thus be either independent or dependent variables.

The following variables are here considered as independent variables to the model.

1. The reactor geometry, the type of diluent used and the reactor pressure p have been decided on at the time of the construction of the plant, and are in that sense independent variables.
   The number of settling legs in service will usually be equal to the total number of legs in order to obtain the highest output.
2. The types of catalyst and cocatalyst have been selected on the basis of product development work.
3. The catalyst activation temperature T_{act}, in case a CrOx catalyst is used. It has been selected in such a way that the target melt index of the product will lie within the range of possible melt index values in the reactor.
4. The reactor temperature T is an independent variable for Ziegler–Natta and metallocene catalysts, because its influence on the melt index is either small or nil, and so its value is kept at a constant value. For CrOx catalysts, it is a dependent variable, as it is used to control the product melt index.
5. The ethylene concentration in the diluent y_2 is an important parameter. There is a certain freedom in the selection of the ethylene concentration. Lower limit is the starving of the catalyst, upper limit is the bubble point of the reactor liquid.
6. The cocatalyst concentration y_{cocat}. It is selected for reasons of catalyst activity, or product quality. A cocatalyst is not always used.
7. The solids concentration C_0 is at first sight controlled by the feed rate of the diluent. However, it has been shown that the solids concentration is not always a univocal function of the diluent feed rate, and that the working point where production rate and removal of solids are in equilibrium is only a metastable equilibrium, see Section 3.3. So it is difficult to explain how operators manage to keep the solids concentration constant at a certain value. The pragmatic approach applied here is to consider this value of the solids concentration as an independent variable.
8. The polymer melt index MFI and density DENS, as measured in the laboratory. They are given as targets to the operator, to be controlled in the reactor. They are thus independent variables to this model.

9. Either the catalyst feed rate F_{cat}, or the production rate P can be considered as an independent variable. The catalyst feed rate will be set at a certain value to obtain a certain production rate. In view of the necessary calculations in the model, it is expedient to adopt the production rate as the independent variable. The catalyst feed rate will follow from it, and is thus listed as a dependent variable.

11.1.3 Dependent Variables

The following variables are considered as being dependent variables.

1. The concentrations of 1-hexene and hydrogen, y_6 and y_H, in the diluent. They are controlled in order to reach the targets of melt index and density at the other selected reactor conditions.
2. The reactor temperature T is a dependent variable as function of the melt index target for chromium oxide catalysts.
3. The polymer density in the reactor ρ_p (to be discerned from the density DENS as measured in the lab) and the liquid density ρ_f.
4. The catalyst activity profile $\overline{A}(\theta, Rx)$ (*kg* polymer per *kg* of catalyst and per unit time) is fixed by the selection of the catalyst and cocatalyst, and the set of reactor conditions Rx.
5. Other polymer properties, apart from melt index and density, like molecular weight distribution, the incorporated ethylene, 1-hexene and hydrogen in the polymer, x_2, x_6, and x_H, the distribution of the comonomer over the chains of the polymer, the powder morphology, and the settling velocity $v_{sett}(Rx)$ are set by the selection of catalyst, cocatalyst, and reactor conditions Rx.
6. The reactor feeds, diluent feed F_{dil}, ethylene feed F_2, hexene feed F_6, hydrogen feed F_H, and cocatalyst feed F_{cocat} follow from the required reactor conditions.
7. The mass of solids in the reactor m_{pol}, the mass of catalyst in the reactor m_{cat}, the catalyst concentration y_{cat}, the catalyst residence time τ.
8. The catalyst reactor activity $\mathbb{A}$, the catalyst reactor productivity $\mathbb{P}$, the catalyst feed rate F_{cat}.

The model will be complete when all dependent variables will have been determined from the independent variables and when this information will be satisfactory and practical to plan a real life reactor run.
The most important dependent variables, i.e. polymer properties and catalyst activities, are determined by the catalyst type and the reactor conditions. They cannot be calculated by deduction, so correlations must be built from practical data, and used in the model to predict these variables by interpolation. The preferred technique to build the correlations is multiple linear regression. Guidelines on how to build correlations are given in Chapter 12.

11.1.4 The Use of Conversions

The calculation of the catalyst reactor activity $\mathbb{A}$ is an important step in the design of the reactor model. However, since up to three reactants, ethylene, 1-hexene and hydrogen, participate in the polymerisation, the catalyst reactor activity $\mathbb{A}$ is the sum of three activities,

$$\mathbb{A} = \mathbb{A}_2 + \mathbb{A}_6 + \mathbb{A}_\mathrm{H} = \sum \mathbb{A}_i ,$$

where $\mathbb{A}_i$ is the catalyst reactor activity for reactant i, or the quantity of reactant i consumed in the polymerisation reaction by all catalyst present in the reactor per unit time, per *kg* catalyst, (*kg* reactant/*kg* catalyst, *s*).

The catalyst reactor activity is a rather complicated function of the catalyst activity profile $\overline{\mathcal{A}}(\theta, Rx)$ and the catalyst residence time distribution, see Chapter 7. Also, it is quite cumbersome to predict the catalyst activity profile for a catalyst, covering a wide range of reactor conditions. Instead of the catalyst reactor activities $\mathbb{A}_i$, conversions will be used in this model, which are more easily obtained from plant reactor data, and are more convenient in the calculations. The conversions of the reactants have been introduced in Chapter 9,

$$\xi_2 = \frac{m_\mathrm{cat}\mathbb{A}_2}{F_2} \; ; \quad \xi_6 = \frac{m_\mathrm{cat}\mathbb{A}_6}{F_6} \; ; \quad \xi_\mathrm{H} = \frac{m_\mathrm{cat}\mathbb{A}_\mathrm{H}}{F_\mathrm{H}} \; .$$

As a combination of a number of already listed variables, they can replace the activity profile $\overline{\mathcal{A}}(\theta, Rx)$, and the reactor activities $\mathbb{A}_i$ in the list of variables.

Since the conversions are relative quantities, they do not suffice to calculate absolute quantities like catalyst activity, catalyst productivity, and catalyst feed rate. However, this is not much to worry about, as conversions allow to define the reactor concentrations, which are the most important parameters to determine the product quality, see Section 11.3.2. The knowledge of catalyst activity or productivity, is only needed to set the catalyst feed rate in order to obtain a certain production rate. This is easily adjusted during a practical reactor run. To predict the catalyst feed rate, one catalyst reactor activity should be used. An empirical correlation for total reactor activity $\mathbb{A}$, or reactor productivity $\mathbb{P}$ is most suited to the purpose.

11.2 Simplifying Assumptions

This simple model cannot function without a number of simplifications. They can be accepted as justified on empirical grounds.

1. The set of *reactor conditions* is represented as,

$$Rx = (T, T_\mathrm{act}, y_2, y_6, y_\mathrm{H}, y_\mathrm{cat}, y_\mathrm{cocat}, C_\mathrm{O}, \rho_\mathrm{p}, \rho_\mathrm{f}) .$$

The concentrations y_i are only measured in one point, i.e. in the flash gas, or the liquid leaving the reactor via the settling legs. It is assumed that this set of single

values of concentrations can be regarded as representative for the concentration profiles in the whole loop reactor, and that one set of measured concentrations in the flash gas represents only one concentration profile in the reactor, irrespective of the production rate or catalyst residence time. Much of this has been corroborated by the discussions in Chapter 10.
A limited set of reactor conditions, $(T, T_{act}, y_2, y_6, y_H, y_{cocat})$, will from here on be understood by the term of *concentration profile*.
2. The polymer properties are set by the selection of the catalyst, cocatalyst, and the concentration profile. It is assumed that the effect of the concentration profile, together with that of the catalyst system, on the polymer properties, is independent of the catalyst residence time and the production rate. This has been learnt during the process of making many correlations, where the production rate has never been a factor influencing the effect of the reactor conditions on catalyst behaviour and polymer properties.
This assumption is more readily acceptable when it is considered that the formation of a polymer chain is effected within a time lapse of a few seconds, which is a small time-scale compared to the catalyst residence time, see Section 5.1.
3. The same experience with correlations has taught that also conversions are function of the concentration profile, but are independent of the catalyst residence time and the production rate.
4. Another simplification has already implicitly been accepted, i.e. that the composition of the reactor liquid is restricted to the components ethylene, comonomer (here 1-hexene), hydrogen and the diluent (isobutane or propane, ...).

11.3 Mathematical Relations

11.3.1 Polymer Properties

The polymer structure needs to be defined in such a way that the polymer properties will suit the final application of the polymer. Only a small part of this is realized in the reactor. In the preparation phase of the reactor run, the optimal catalyst type has already been selected to produce a suitable polymer at a certain value of melt index and density. All that can be done in the reactor, is to control the melt index (MFI) and density (DENS) at the prescribed level. This reduces the number of polymer properties featured in the model to just those two.

Correlations for MFI and DENS produced by the selected type of catalyst system should be made from either lab scale data, or, preferably, from full scale reactor data. MFI and DENS are often correlated with ratios of concentrations, 1-hexene and hydrogen to ethylene, in the reactor liquid. This is highly imprecise, because the ratios of concentrations depend on the conversions of the reactants, and those are not constant. The best correlations are built by relating MFI and DENS as directly as possible to the composition of the polymer, and that is to the incorporated 1-hexene and hydrogen.

The density is, for all catalyst families, see Section 9.5.2, related to the incorporated hexene and the melt index as,

$$\mathrm{DENS} = \beta_0 + \beta_1 (x_6)^{\alpha} + \beta_2 \log(\mathrm{MFI}) \, .$$

The incorporated 1-hexene x_6, is immediately obtained from this correlation, where DENS and MFI are known as the target values.

The melt index, for a Ziegler–Natta catalyst, is related to the incorporated hydrogen and hexene, and to the reactor temperature, see Section 9.5.1,

$$\log(\mathrm{MFI}) = \beta_0 + \beta_1 x_\mathrm{H} + \beta_2 (x_\mathrm{H})^2 + \beta_3 (x_6) + \beta_4 T \, .$$

The incorporated hydrogen x_H, is obtained from this correlation.

For a CrOx catalyst, the reactor temperature T is used to control the MFI, and the activation temperature is another important independent variable. Hydrogen is seldom used. Here it is indicated to choose a certain value for x_H, (make it an independent variable), which is usually zero. The reactor temperature T follows then from the melt index target.

11.3.2 Conversions

The conversions are function of the concentration profile in the reactor. Especially the ethylene concentration has the greatest impact on the conversions. Correlations should be built for each of the three conversions,

$$\xi_i = f(T, T_\mathrm{act}, y_2, y_6, y_\mathrm{H}, y_\mathrm{cocat}) \, , \tag{11.1}$$

wherefrom the three conversions can be calculated. See Sections 9.2, 9.3 and 9.4.
The ethylene concentration is considered as an independent variable, and thus known. The 1-hexene and hydrogen concentrations are now calculated as follows.
The fractions of incorporated reactants are,

$$x_6 = \frac{F_6 \xi_6}{P} \, ; \quad x_\mathrm{H} = \frac{F_\mathrm{H} \xi_\mathrm{H}}{P} \, ; \quad x_2 = 1 - x_6 - x_\mathrm{H} = \frac{F_2 \xi_2}{P} \, .$$

Whence the ratios of 1-hexene and hydrogen to ethylene in the feeds are calculated,

$$\frac{F_6}{F_2} = \frac{\xi_2}{\xi_6} \frac{x_6}{1 - x_6 - x_\mathrm{H}} \, ; \quad \frac{F_\mathrm{H}}{F_2} = \frac{\xi_2}{\xi_\mathrm{H}} \frac{x_\mathrm{H}}{1 - x_6 - x_\mathrm{H}} \, .$$

The concentrations of the reactants in the reactor liquid are equal to their concentrations in the flash gas,

$$y_2 = \frac{F_2 (1 - \xi_2)}{D_\mathrm{R}} \frac{(\mathrm{MW})_\mathrm{fg}}{28.05} \, ; \quad y_6 = \frac{F_6 (1 - \xi_6)}{D_\mathrm{R}} \frac{(\mathrm{MW})_\mathrm{fg}}{84.16} \, ; \quad y_\mathrm{H} = \frac{F_\mathrm{H} (1 - \xi_\mathrm{H})}{D_\mathrm{R}} \frac{(\mathrm{MW})_\mathrm{fg}}{2.01} \, .$$

Combining the expressions above, the 1-hexene and hydrogen concentrations in the liquid can now be calculated as a function of the ethylene concentration,

$$y_6 = y_2 \frac{\xi_2}{\xi_6} \frac{1-\xi_6}{1-\xi_2} \frac{x_6}{1-x_6-x_H} \frac{28.05}{84.16} ; \tag{11.2}$$

$$y_H = y_2 \frac{\xi_2}{\xi_H} \frac{1-\xi_H}{1-\xi_2} \frac{x_H}{1-x_6-x_H} \frac{28.05}{2.01} . \tag{11.3}$$

A number of iteration steps is necessary, starting from a certain starting value for the concentrations of 1-hexene and hydrogen in the liquid, since the conversions are supposed to be a function of these concentrations, and making use of equations (11.1), (11.2), and (11.3).

No iteration is needed when the conversions are represented satisfactorily by correlations independent of the 1-hexene and hydrogen concentrations, like is the case in this example for a Ziegler–Natta catalyst,

$$\xi_2 = \beta_0 + \beta_1 y_2 + \beta_2 y_{\text{cocat}} + \beta_3 T ,$$
$$\xi_6 = \beta_0 + \beta_1 y_2 + \beta_2 y_{\text{cocat}} + \beta_3 T ,$$
$$\xi_H = \beta_0 + \beta_1 y_2 + \beta_2 (y_2)^2 + \beta_3 y_{\text{cocat}} .$$

For a metallocene catalyst, the hydrogen conversion is even simply equal to 100% !

The model is now already capable to steer the quality (melt index and density) of the polymer, since the conversions, the ratios in the flash gas and the ratios in the feeds are known. To control the product quality, the ethylene concentration should be kept as constant as possible, because of its great impact on the conversions. It is recommended to use the ratios in the feeds, F_H/F_2 and F_6/F_2, to steer the quality, since this will allow a constant ethylene concentration more easily, and this *steering* of the feeds is more direct and fast than the *controlling* of ratios in the flash gas, which are only known with a considerable time delay.

The selection of the set point for the ethylene concentration is important, because it influences the absolute values of the hexene and hydrogen concentrations in the liquid needed to produce a certain quality. A proper value of the ethylene concentration can keep the reactor concentration profile away from its limits imposed by e.g. bubble point or fouling curve, see Section 9.7.

11.3.3 Physical Properties

The density of the polymer in the reactor ρ_p is taken equal to the density measured in the lab at room temperature DENS, and corrected for the thermal expansion at the reactor temperature T. No “swelling” due to the absorption of liquid is taken into account. The porosity of the HDPE particles is very low. A correlation describing the polymer density as a function of the temperature is used, of the type,

$$\rho_p = \text{DENS} + \beta_0 + \beta_1 T + \beta_2 T^2 .$$

The density of the liquid in the reactor, ρ_f, can be calculated from an equation of state, or from a correlation built on a large number of such calculations covering the range of concentrations in which the reactor is operated, and taking into account the reactor pressure and the type of diluent used,

$$\rho_f = f(T, p, y_2, y_6, y_H) = \beta_0 + \beta_1 y_2 + \beta_2 y_6 + \beta_3 y_H + \beta_4 T \, .$$

The average molecular weight of the reactor liquid (or flash gas) is calculated as,

$$(\mathrm{MW})_{fg} = 28.05 y_2 + 84.16 y_6 + 2.01 y_H + (1 - y_2 - y_6 - y_H)(\mathrm{MW})_{dil} \, .$$

11.3.4 Catalyst Activity

The mass of polymer in the reactor is,

$$m_{pol} = V_R C_O \rho_p \, .$$

The average catalyst residence time is,

$$\tau = \frac{V_R C_O \rho_p}{P} \, .$$

To determine the catalyst productivity, a correlation is needed based on experimental data, either for the catalyst reactor productivity, or for the reactor activity,

$$\mathbb{P} = f(\tau, T, T_{act}, y_2, y_6, y_H, y_{cocat}) \, ,$$

$$\mathbb{A} = \frac{\mathbb{P}}{\tau} \, .$$

The catalyst feed is,

$$F_{cat} = \frac{P}{\mathbb{P}} \, .$$

The mass of catalyst in the reactor is,

$$m_{cat} = \tau F_{cat} = \frac{V_R C_O \rho_p}{\mathbb{P}} \, .$$

11.3.5 Flash Gas Flows

In an HDPE loop reactor, equipped with settling legs, the total flash gas flow is strongly influenced by the settling of the polymer particles in the settling legs. A correlation from empirical data is needed to determine the settling velocity of the polymer particles in the settling legs, as described in Section 3.2.2,

$$v_{sett} = \beta_0 - \beta_1 C_O + \beta_2 \frac{\rho_p - \rho_f}{\rho_f} \, .$$

Then the total flow of recycled diluent or flash gas D_R can be calculated from the solids concentration control equation (3.12) obtained in Chapter 3 on the functioning of the settling legs,

$$P = \frac{\rho_p C_O}{1 - C_O}\left(\frac{D_R}{\rho_f} + \frac{\pi D^2}{4} v_{sett} N\right),$$

whence,

$$D_R = \rho_f\left(P\frac{1 - C_O}{\rho_p C_O} - \frac{\pi D^2}{4} v_{sett} N\right).$$

The flash gas flow is limited by the circulation capacity of the recovery system, and may be one of the bottlenecks to the production capacity when the settling velocity is low. The flows of the reactants ethylene, 1-hexene and hydrogen in the flash gas are,

$$(FG)_i = \frac{D_R}{(MW)_{fg}} y_i (MW)_i .$$

11.3.6 Reactor Feeds

The reactor feeds are,

$$F_2 = P\frac{1 - x_6 - x_H}{\xi_2};$$

$$F_6 = P\frac{x_6}{\xi_6};$$

$$F_H = P\frac{x_H}{\xi_H};$$

$$F_{dil} = \frac{D_R}{(MW)_{fg}}(1 - y_2 - y_6 - y_H)(MW)_{dil} .$$

11.3.7 Constraints

All important parameters have been treated. To make the model complete, it is recommended to check whether a number of constraints have not been infringed, like,

1. limitations to the operation of the reactor, like bubble point of the liquid and fouling curve;
2. limitations to the heat transfer capacity from the reactor to the cooling jackets, and of the reactor coolant cooler; if very high production rates are to be achieved, then it is useful to build a model that describes the cooling system of the reactor as a function of the production rate;
3. limitation on the reactor circulation pump power, usually set by the maximal volumetric solids concentration;
4. limitations to the raw material feed rates, upper limits as well as lower limits (at which the flow can still be controlled);
5. limitations to the circulation capacity of the flash gas and recovery systems.

11.4 A Corrective Off-line Model

The model as it has been described here, will be of great help to prepare runs of new products. It will also be useful to assist the operator when corrective steps should be taken during the reactor run itself, to control melt index and density.

When a transition is made to another grade, or when melt index and density are far off target, a rather large correction should be made to the reactor conditions. A trap to be avoided at all cost then, is to handle the system as an ordinary control loop, where the response to the input is fast, and to make a long series of small corrections, responding to every new analysis of melt index or density. This does not work because the system is slow. The operator will often not see the result of his input before his shift time is over ...

It is a much better strategy to allow the reactor to stabilise, and await the results under stable conditions. Action should be taken when a period of at least three consecutive hours of stable operation has been realized. The correlations for melt index, density and conversions are updated with the information from that stable period, and the model is applied, using the updated correlations.

The average values within that period should be recorded for melt index, density, and those variables that are necessary to calculate the conversions of the reactants, and to predict melt index and density from the correlations used in the model. Let these values be called *real-time values*.

The real-time conversions are calculated, according to the equations (9.4), (9.7), and (9.9),

$$\xi_2 = 1 - \frac{F_{\mathrm{IC4}}}{F_2}\frac{y_2}{y_{\mathrm{IC4}}}\frac{28.05}{58.12}\,; \quad \xi_{\mathrm{H}} = 1 - \frac{F_{\mathrm{IC4}}}{F_{\mathrm{H}}}\frac{y_{\mathrm{H}}}{y_{\mathrm{IC4}}}\frac{2.01}{58.12}\,; \quad \xi_6 = 1 - \frac{F_{\mathrm{IC4}}}{F_6}\frac{y_6}{y_{\mathrm{IC4}}}\frac{84.16}{58.12}\,;$$

as well as the real-time incorporated hydrogen and 1-hexene, equations (9.10) and (9.11),

$$x_{\mathrm{H}} = \frac{F_{\mathrm{H}}\xi_{\mathrm{H}}}{F_2\xi_2 + F_6\xi_6 + F_{\mathrm{H}}\xi_{\mathrm{H}}}\,; \quad x_6 = \frac{F_6\xi_6}{F_2\xi_2 + F_6\xi_6 + F_{\mathrm{H}}\xi_{\mathrm{H}}}\,.$$

These real-time values are used as input to the correlations for melt index and density, Section 11.3.1, and for the conversions, equation (11.1). The obtained values are the *predicted values* for melt index, density and conversions, based on the real-time data,

$$\begin{aligned}
\log(\widehat{\mathrm{MFI}}) &= f(T, T_{\mathrm{act}}, x_{\mathrm{H}}, x_6)\,,\\
\widehat{\mathrm{DENS}} &= \beta_0 + \beta_1(x_6)^{\alpha} + \beta_2\log(\mathrm{MFI})\,,\\
\widehat{\xi_i} &= f(T, T_{\mathrm{act}}, y_2, y_6, y_{\mathrm{H}})\,.
\end{aligned}$$

Update constants, k, to the correlations, are found as the differences between the predicted values and the real time values,

$$k_{\mathrm{MFI}} = \log(\mathrm{MFI}) - \log(\widehat{\mathrm{MFI}}) ,$$
$$k_{\mathrm{DENS}} = \mathrm{DENS} - \widehat{\mathrm{DENS}} ,$$
$$k_{\xi,i} = \xi_i - \widehat{\xi_i} .$$

The correlations are updated with these constants, which means that their predicted values are simply shifted to match the level of the real-time values of melt index, density and conversions,

$$\log(\mathrm{MFI}) = f(T, T_{\mathrm{act}}, x_{\mathrm{H}}, x_6) + k_{\mathrm{MFI}} ,$$
$$\mathrm{DENS} = \beta_0 + \beta_1(x_6)^{\alpha} + \beta_2 \log(\mathrm{MFI}) + k_{\mathrm{DENS}} ,$$
$$\xi_i = f(T, T_{\mathrm{act}}, y_2, y_6, y_{\mathrm{H}}) + k_{\xi,i} .$$

The model is now applied, as described in the Sections 11.3.1 and 11.3.2, using the updated correlations, to calculate the ratios in the feeds F_{H}/F_2 and F_6/F_2 necessary to reach the new targets for melt index and density.

Computer programs and models should never be used without proper judgement. A critical look should be taken at each update constant k before applying it. Is it insignificant, or significant, or suspect? To judge its value, it is helpful to express it as a multiple of the standard error of the estimate (SEE) of the correlation it is supposed to improve. When applied, it is good to provide a filter factor, between 0 and 1, to multiply the update constant with, and allow the user to express his confidence.

Notations

The reactants and chemical substances are indicated by a subscript:

2	ethylene
6	1-hexene
H	hydrogen
i	one of the reactants ethylene, 1-hexene or hydrogen
IC4	isobutane
cat	catalyst
cocat	cocatalyst

$\mathbb{A}$	catalyst reactor activity: quantity of polymer produced by all catalyst present in the reactor per unit time, per *kg* catalyst, (*kg* polymer/*kg* catalyst, *s*).
$\mathbb{A}_2$	catalyst reactor activity for ethylene: quantity of ethylene consumed in the polymerisation reaction by all catalyst present in the reactor per unit time, per *kg* catalyst, (*kg* ethylene/*kg* catalyst, *s*).

$\mathbb{A}_6$	catalyst reactor activity for 1-hexene: quantity of 1-hexene consumed in the polymerisation reaction by all catalyst present in the reactor per unit time, per *kg* catalyst, (*kg* 1-hexene/*kg* catalyst, *s*).
$\mathbb{A}_H$	catalyst reactor activity for hydrogen: quantity of hydrogen consumed in the polymerisation reaction by all catalyst present in the reactor per unit time, per *kg* catalyst, (*kg* hydrogen/*kg* catalyst, *s*).
$\overline{\mathcal{A}}(\theta, Rx)$	catalyst activity profile: quantity of polymer produced per *kg* of catalyst, per unit time, at a certain lifetime $t = \theta$ in the reactor, (*kg* polymer/*kg* catalyst, *s*).
C_0	volumetric concentration of solids in the reactor, (–).
D	settling leg internal diameter, (*m*).
DENS	density of the polymer measured in the lab at room temperature, (kg, m^{-3}).
D_R	flow of recycled diluent (reactor liquid, or flash gas) leaving the reactor, (kg, s^{-1}).
F_2	ethylene feed to the reactor, (kg, s^{-1}).
F_6	1-hexene feed to the reactor, (kg, s^{-1}).
F_H	hydrogen feed to the reactor, (kg, s^{-1}).
F_{cat}	catalyst feed to the reactor, (kg, s^{-1}).
F_{cocat}	cocatalyst feed to the reactor, (kg, s^{-1}).
F_{dil}	diluent (isobutane, propane, ...) feed to the reactor, (kg, s^{-1}).
$(FG)_i$	flow of reactant *i* in the flash gas, (kg, s^{-1}).
k	update constant to a correlation, equal to the difference between the predicted value and the real time value.
m_{cat}	mass of catalyst present in the reactor, (*kg*).
m_{pol}	mass of polymer present in the reactor, (*kg*).
MFI	melt index of the polymer, (dg, min^{-1}).
$(MW)_{fg}$	average molecular weight of the reactor liquid (or flash gas), leaving the reactor, ($kg, kmole^{-1}$).
$(MW)_{dil}$	molecular weight of the diluent (isobutane or propane or ...), ($kg, kmole^{-1}$).
N	number of settling legs in service, (–).
p	reactor pressure, (*Pa*).
P	polymer production rate: total quantity of polymer produced in the reactor per unit of time, (*kg* polymer, s^{-1}).
$\mathbb{P}$	catalyst reactor productivity: total quantity of polymer produced by one *kg* of catalyst that has left the reactor, (*kg* polymer/*kg* catalyst).
Rx	symbol representing all of the reactor conditions, including temperature, concentrations, flows, and densities.
T	reactor temperature, (°C).
T_{act}	chromium oxide catalyst activation temperature, (°C).
V_R	reactor volume, (m^3).

v_{sett}	effective settling velocity by which settling polymer particles enter the settling leg, (m, s^{-1}).
x_2	incorporated ethylene in the polymer, (*kg* ethylene/*kg* polymer).
x_6	incorporated 1-hexene in the polymer, (*kg* 1-hexene/*kg* polymer).
x_H	incorporated hydrogen in the polymer, (*kg* hydrogen/*kg* polymer).
y_2	ethylene concentration, (*mole* ethylene/*mole* reactor liquid).
y_6	1-hexene concentration, (*mole* 1-hexene/*mole* reactor liquid).
y_H	hydrogen concentration, (*mole* hydrogen/*mole* reactor liquid).
y_{cat}	catalyst concentration, (*kg* catalyst/*kg* reactor liquid).
y_{cocat}	cocatalyst concentration, (*kg* cocatalyst/*kg* reactor liquid).
ξ_2	ethylene conversion in the reactor: the fraction of all ethylene that is fed to the reactor, which is consumed in the polymerisation reaction, (–).
ξ_6	1-hexene conversion in the reactor: the fraction of all 1-hexene that is fed to the reactor, which is consumed in the polymerisation reaction, (–).
ξ_H	hydrogen conversion in the reactor: the fraction of all hydrogen that is fed to the reactor, which is consumed in the polymerisation reaction, (–).
ρ_f	density of the liquid in the reactor, (kg, m^{-3}).
ρ_p	density of the polymer in the reactor, (kg, m^{-3}).
τ	average catalyst residence time, (s).

12 Establishing Correlations from Loop Reactor Data by Linear Regression

It has been demonstrated in this book that experimental data is vital to understand the functioning of the slurry loop reactor, and to construct models to control it. The way to obtain insight into this data, and to transform it into tools for practical use, is to establish correlations between the variables. The technique of multiple linear regression is most appropriate to treat the reactor data, be it from full scale operation or from bench scale (lab) reactors.

Linear regression is described and discussed in other books and courses by persons skilled in the art. The following text is not meant to replace them, but to share some experience on how linear regression is applied in this field. The technique is so important that a full chapter is dedicated to it, even if this matter can hardly be qualified as reactor technology. Too often correlations appear that are quite useless, either because they are built on insufficient data, or because they are obtained by a thoughtless application of some computer program.

The methods mentioned here are available in easy statistical software programs, that are not meant for the professional statistician.

12.1 The Data Set

No valuable correlation can be set up without a valuable and balanced data set. To construct and interpret a data set, it is necessary to have a profound and basic idea of what the construction of a data file should be like. The rather abstract representation and treatment of the data that follows here, is not intended to be practical, but it should be kept in mind when interpreting the data. In the next sections, the ideas will be applied in practical proceedings that are far less complicated.

A data set presents itself as a table composed of a number of columns and rows, where each column contains the values of one variable, and each row contains the values of the variables from one test, data point or *observation*. Such a row may be a selection of hourly averages of reactor parameters at a specific hour, taken from a process computer data base, or it may be the values of the parameters of a single bench scale polymerisation test.

Distinction should be made between the *independent variables* X_i, which can be freely chosen and determine the reactor conditions, and the *dependent variables* Y, which are the result of the selected reactor conditions, and which should be related to those reactor conditions.

Let the p independent variables $X_1 \ldots X_i \ldots X_p$ be considered. The table of their values x_{ji} forms a matrix of p columns and n rows, when there are n observations t_j,

$$\begin{array}{c|ccccc} & X_1 & \ldots & X_i & \ldots & X_p \\ \hline t_1 & x_{11} & \ldots & x_{1i} & \ldots & x_{1p} \\ \vdots & \vdots & \ddots & \vdots & \ddots & \vdots \\ t_j & x_{j1} & \ldots & x_{ji} & \ldots & x_{jp} \\ \vdots & \vdots & \ddots & \vdots & \ddots & \vdots \\ t_n & x_{n1} & \ldots & x_{ni} & \ldots & x_{np} \end{array} .$$

In a first step, the range of each variable should be compared to the usual operating window of the process, and should be found to cover an important part of that window. It makes no sense to consider a variable that is more or less constant. Such a variable should better be excluded from the calculations, but this should be mentioned when reporting the results.

The absolute values of the (independent) variables can vary a lot, from mole fractions of 0.01 *mole*/*kmole* to temperatures of 100 °C and up to catalyst productivities of several thousands of *kg*/*kg*. They also depend on the units chosen to express their value. Therefore, to make them comparable, all values are standardised so that their range is stretching from -1.0 to $+1.0$. If the highest value of X_i is equal to u_i, and the lowest value of X_i is equal to l_i, then every value x_{ji} is replaced by,

$$a_{ji} = \frac{x_{ji} - \frac{1}{2}(u_i + l_i)}{\frac{1}{2}(u_i - l_i)} , \tag{12.1}$$

with $-1 < a_{ji} < +1$.
A matrix $\underline{A}_{n,p}$ of the standardised values of the independent variables is thus formed,

$$\underline{A}_{n,p} = \begin{vmatrix} a_{11} & \ldots & a_{1i} & \ldots & a_{1p} \\ \vdots & \ddots & \vdots & \ddots & \vdots \\ a_{j1} & \ldots & a_{ji} & \ldots & a_{jp} \\ \vdots & \ddots & \vdots & \ddots & \vdots \\ a_{n1} & \ldots & a_{ni} & \ldots & a_{np} \end{vmatrix} .$$

Each observation is represented by one *row* of the matrix $\underline{A}_{n,p}$, and can be considered as a vector,

$$\overline{a}_j = (a_{j1}, \ldots, a_{ji}, \ldots, a_{jp}) ,$$

in a vector space of p dimensions, where the dimensions are the independent variables. There are n such *observation vectors*.

The matrix $\underline{A}_{n,p}$ is now multiplied at the right side by its transposed matrix $\underline{A}^{\mathrm{T}}_{p,n}$. The matrix product is a square $n \times n$ matrix. Each element (j, k) of this *matrix of observations* is the dot product of two observation vectors $\overline{a}_j \cdot \overline{a}_k$,

$$\underline{A}_{n,p} \cdot \underline{A}^{\mathrm{T}}_{p,n} = \begin{vmatrix} a_{11} & \dots & a_{1i} & \dots & a_{1p} \\ \vdots & \ddots & \vdots & \ddots & \vdots \\ a_{j1} & \dots & a_{ji} & \dots & a_{jp} \\ \vdots & \ddots & \vdots & \ddots & \vdots \\ a_{n1} & \dots & a_{ni} & \dots & a_{np} \end{vmatrix} \cdot \begin{vmatrix} a_{11} & \dots & a_{j1} & \dots & a_{n1} \\ \vdots & \ddots & \vdots & \ddots & \vdots \\ a_{1i} & \dots & a_{ji} & \dots & a_{ni} \\ \vdots & \ddots & \vdots & \ddots & \vdots \\ a_{1p} & \dots & a_{jp} & \dots & a_{np} \end{vmatrix} ,$$

or,

$$\left(\underline{A} \cdot \underline{A}^{\mathrm{T}}\right)_{n,n} = \begin{vmatrix} \overline{a}_1 \cdot \overline{a}_1 & \dots & \overline{a}_1 \cdot \overline{a}_j & \dots & \overline{a}_1 \cdot \overline{a}_n \\ \vdots & \ddots & \vdots & \ddots & \vdots \\ \overline{a}_j \cdot \overline{a}_1 & \dots & \overline{a}_j \cdot \overline{a}_j & \dots & \overline{a}_j \cdot \overline{a}_n \\ \vdots & \ddots & \vdots & \ddots & \vdots \\ \overline{a}_n \cdot \overline{a}_1 & \dots & \overline{a}_n \cdot \overline{a}_j & \dots & \overline{a}_n \cdot \overline{a}_n \end{vmatrix} .$$

The dot product of two vectors is,

$$\overline{a}_j \cdot \overline{a}_k = |\overline{a}_j|\,|\overline{a}_k| \cos\theta_{jk} ,$$

where θ_{jk} is the angle between the two vectors. Hence that on the diagonal of the matrix $(\underline{A} \cdot \underline{A}^{\mathrm{T}})_{n,n}$ the squares of the norms of the vectors $\overline{a}_j$ are found, and also that the matrix is symmetric since the dot product of two vectors is commutative.
When each element (j, k) of this matrix of observations is divided by the norms of the two vectors $\overline{a}_j$ and $\overline{a}_k$, that is by $|\overline{a}_j|\,|\overline{a}_k|$, a matrix $(\cos\theta_{jk})$ is obtained of which each element (j, k) is equal to the cosine of the angle between two observation vectors $\overline{a}_j$ and $\overline{a}_k$.

Each observation can be thus be considered as a vector in the space of the independent variables. The matrices $(\underline{A} \cdot \underline{A}^{\mathrm{T}})_{n,n}$ and $(\cos\theta_{jk})$ show, in a condensed form, the norms of the vectors and the angle between each pair of vectors. Two observations may own nearly the same direction when $\cos\theta_{jk} \to +1$. If on top of that their norms are nearly equal, then the two observations are nearly identical.

There is no harm in some observations showing the same direction, but, if too many observations do so, there is a risk that the independent variables be correlated.

There is no harm either in having a number of nearly identical observations. Actually, repeating one observation a number of times will help to reduce the noise in the data, and that is exactly the advantage of the use of statistical methods. However, if a data set contains a larger group of nearly identical observations, that group may have a too great influence on the correlation, as the calculations will try and reduce the variance within that group, which, perhaps, may be just noise. A well balanced data set should comprise more or less equal numbers of repeated observations.

The calculation of a linear correlation is very similar to the solving of a set of simultaneous linear equations, where the unknowns are the correlation coefficients plus the constant term. Indeed, if linear regression is applied to a small set of n observations, which in number equals the number of independent variables plus one, $n = p + 1$, the obtained correlation will be the exact solution of a system of n linear equations. However meaningless the result may be, the Pearson correlation coefficient will be equal to 1.00, (R-squared equal to 100%), as for the perfect correlation. The more the number of observations is exceeding $p + 1$, the better the chances are that the correlation will cover the process window and produce meaningful results. But one is mistaken when being happy with a large data set where actually most observations are just repeating the same reactor conditions, and do not bring extra information. Therefore, it is recommended to consider each group of repeated observations as one single observation, and to keep the number of correlation coefficients plus 1 smaller than the number of groups of repeated observations, lest the correlation will resemble the solution of a system of equations and in fact be quite meaningless. This reasoning also explains why striving for high values of R-squared by adding more and more independent variables to the correlation leads to an illusion of accuracy, and unrealistic results.

After concentrating on the observations, the independent variables are now considered. Each *column* of the matrix $\underline{A}_{n,p}$, comprises the values of one independent variable, and can also be considered as a vector,

$$\overline{v}_i = (a_{1i}, \ldots, a_{ji}, \ldots, a_{ni}) ,$$

in a vector space of n dimensions, where the dimensions are the observations. There are p such *variable vectors*.
It is more difficult to represent a variable vector to the mind than an observation vector. Its n components are the standardised values of the variable X_i in the n observations t_j. Each of the n observations is one of the n dimensions. It seems strange that in one dimension, or observation, the components of the different variable vectors originate from values that were expressed in different units, and thus were not comparable. However, the standardised values of the variables are dimensionless, and the goal of it all is to investigate whether or not variables show similar evolutions in their relative magnitude from observation to observation.

The matrix $\underline{A}_{n,p}$ is now multiplied at the left side by its transposed matrix $\underline{A}^{\mathrm{T}}_{p,n}$. The matrix product is a square $p \times p$ matrix. Each element (i, l) of this *matrix of variables* is the dot product of two independent variable vectors $\overline{v}_i \cdot \overline{v}_l$,

$$\underline{A}^{\mathrm{T}}_{p,n} \cdot \underline{A}_{n,p} = (\underline{A}^{\mathrm{T}} \cdot \underline{A})_{p,p} = \begin{vmatrix} \overline{v}_1 \cdot \overline{v}_1 & \ldots & \overline{v}_1 \cdot \overline{v}_i & \ldots & \overline{v}_1 \cdot \overline{v}_p \\ \vdots & \ddots & \vdots & \ddots & \vdots \\ \overline{v}_i \cdot \overline{v}_1 & \ldots & \overline{v}_i \cdot \overline{v}_i & \ldots & \overline{v}_i \cdot \overline{v}_p \\ \vdots & \ddots & \vdots & \ddots & \vdots \\ \overline{v}_p \cdot \overline{v}_1 & \ldots & \overline{v}_p \cdot \overline{v}_i & \ldots & \overline{v}_p \cdot \overline{v}_p \end{vmatrix} .$$

On the diagonal of the matrix $(\underline{A}^{\mathrm{T}} \cdot \underline{A})_{p,p}$ the squares of the norms of the vectors $\overline{v}_i$ are found, and the matrix is symmetric.
When each element (i, l) of this matrix of variables is divided by the norms of the two vectors $\overline{v}_i$ and $\overline{v}_l$, a matrix $(\cos\theta_{il})$ is obtained of which each element (i, l) is equal to the cosine of the angle between two variable vectors $\overline{v}_i$ and $\overline{v}_l$.

The variable vectors are *orthogonal* when the matrix $(\cos\theta_{il})$ is equal to the unit matrix $\underline{E}_{p,p}$, of which the elements on the diagonal are equal to +1, and all the other elements are zero[1]. It means that the variable vectors are perpendicular to each other, or that the variables are completely independent of each other, or, not correlated.
Indeed, considering the matrix $(\cos\theta_{il})$ in more detail,

$$\cos\theta_{il} = \frac{\overline{v}_i \cdot \overline{v}_l}{|\overline{v}_i|\,|\overline{v}_l|},$$

$$\cos\theta_{il} = \frac{a_{1i}a_{1l} + \cdots + a_{ji}a_{jl} + \cdots + a_{ni}a_{nl}}{(a_{1i}^2 + \cdots + a_{ji}^2 + \cdots + a_{ni}^2)^{1/2}(a_{1l}^2 + \cdots + a_{jl}^2 + \cdots + a_{nl}^2)^{1/2}},$$

$$\cos\theta_{il} = \frac{\sum_j a_{ji}a_{jl}}{\left(\sum_j a_{ji}^2\right)^{1/2}\left(\sum_j a_{jl}^2\right)^{1/2}}.$$

Turning back to non standardised variables, using equation (12.1),

$$\cos\theta_{il} = \frac{\sum_j [x_{ji} - \frac{1}{2}(u_i + l_i)][x_{jl} - \frac{1}{2}(u_l + l_l)]}{\left\{\sum_j [x_{ji} - \frac{1}{2}(u_i + l_i)]^2\right\}^{1/2}\left\{\sum_j [x_{jl} - \frac{1}{2}(u_l + l_l)]^2\right\}^{1/2}}.$$

When the data points in the data set are well spread and free of extreme outliers, then $(1/2)(u_i + l_i)$ is approximately equal to the mean $\overline{x}_i$ of the variable X_i, and,

$$\cos\theta_{il} \approx \frac{\sum_j (x_{ji} - \overline{x}_i)(x_{jl} - \overline{x}_l)}{\left[\sum_j (x_{ji} - \overline{x}_i)^2\right]^{1/2}\left[\sum_j (x_{jl} - \overline{x}_l)^2\right]^{1/2}}.$$

The right hand member of the equation is recognised as the product-moment or Pearson correlation coefficient of the two variables X_i and X_l. So the elements of the matrix $(\cos\theta_{il})$ are approximately equal to the Pearson correlation coefficients of the variables X_i and X_l, $\cos\theta_{il} \approx R_{il}$. When the variable vectors are orthogonal, they are not correlated at all.

In the ideal data set, the variable vectors are orthogonal. It means that the independent variables are really independent of each other, and that the effect of each separate variable can be calculated. When two variables are highly correlated, it is not possible to discriminate between the effects of the two.
When all observations are repeated the same number of times, the orthogonality of the variables is conserved: the number of variables is not changed, and the relative directions of the variable vectors are conserved.

---

1 A matrix $\underline{A}$ is said to be orthogonal when $\underline{A} \cdot \underline{A}^{\mathrm{T}}$ is equal to a constant times the unit matrix $\underline{E}$.

Unlike observation vectors, couples of variable vectors must not own opposite directions, ($\cos\theta_{jl} \to -1$), because also then they are completely correlated, albeit negatively.

Example
The following fictitious small data set reflects a designed experiment with 2 independent (standardised) variables and 4 observations (2^2 factorial design),

	X_1	X_2
t_1	−1	+1
t_2	+1	+1
t_3	−1	−1
t_4	+1	−1

.

The matrix of observations is,

$$\underline{A}_{4,2} \cdot \underline{A}^{\mathrm{T}}_{2,4} = (\underline{A} \cdot \underline{A}^{\mathrm{T}})_{4,4} = 2 \cdot \begin{vmatrix} +1 & 0 & 0 & -1 \\ 0 & +1 & -1 & 0 \\ 0 & -1 & +1 & 0 \\ -1 & 0 & 0 & +1 \end{vmatrix} .$$

The 4 observations are either perpendicular or show opposite directions.
The matrix of variables is,

$$\underline{A}^{\mathrm{T}}_{2,4} \cdot \underline{A}_{4,2} = (\underline{A} \cdot \underline{A}^{\mathrm{T}})_{2,2} = 4 \cdot \begin{vmatrix} +1 & 0 \\ 0 & +1 \end{vmatrix} .$$

The two variables are perfectly orthogonal, and thus not correlated. □

12.2 Building the Data Set

When collecting data from full scale reactor operation, one observation, or data point, will consist of a series of hourly averages of the important reactor and product parameters, taken at the same production time, during a stable period, and extracted from a process computer. It is recommended to establish a standard list of variables that should be recorded for each data point. The list should comprise all variables that define the reactor conditions and the product. A number of calculated variables, which are calculated from the recorded variables, e.g. the conversions or the fluff settling velocity, is added to the list. It is very helpful to read the variables in a strict and constant order, so that standard files can be used, and files can easily be joined together.

When collecting data from *laboratory polymerisations*, one has the great advantage that the tests can be designed in number and in choice of parameters. On the other hand, the number of tests is limited due to the work and time involved to get a lab scale reactor running and effect the tests. The number of available data points from a *full scale loop reactor* is almost infinite, but the choice of data points is imposed by the production planning, which leaves little room for experimentation. In order to establish a data set that is complete enough, the data points should be collected for a longer time, during which the building up of the data set is monitored by fitting the available data points into a factorial design of experiments scheme, in order to check the orthogonality of the variables. Each commercial grade, which is produced at standard reactor conditions, should be regarded as one data point in the scheme. Several grades are usually produced from a single Ziegler–Natta or metallocene catalyst, helping to dress the test scheme. A chromium oxide catalyst, however, will produce only one or just a few grades, but has the advantage that the effects of the reactor parameters are the same for all chromium oxide catalysts, so that the data can be massed together, at the condition that each catalyst type is seen as an additional variable.

Still, filling a scheme with data from the current commercial grades will be disappointing. Advantage should be taken of all test runs and abnormal conditions to record extra data points. The periods of stable operation will there be short, but the data should nevertheless be screened carefully for stability. Unstable data points should be eliminated, even if this makes the number of data points still smaller. It is useless to collect a large number of data points that are not valuable.

But to compose an effectively balanced and somewhat orthogonal data set, it will be necessary to organise tests on purpose, where variation is brought into the reactor conditions during the production of the same grade. The reactor conditions to be varied are those that influence the conversions of the reactants, like ethylene concentration, temperature, or cocatalyst dosage. It may sound like heresy to vary reactor conditions in the production of standard grades, but the experienced production engineer will know which grades are less critical to the customer, and thus allow some experimentation, and how far the variations can be pushed without harming the product quality.

Special attention should be paid to data from homopolymer production, where no comonomer is fed. The comonomer feed flow measurement and flash gas concentration will probably not show zero, but some low values that are just noise and thus meaningless. Because properties like polymer density are strongly affected by low concentrations of the comonomer, it is, in such cases, highly recommended to put all comonomer flows, concentrations and incorporated comonomer equal to a neat zero. If not, models would attach great weight to meaningless low concentrations, and as the operator will put in a zero for comonomer feeds and concentrations when using the model for homopolymers, large deviations would be introduced. The same procedure is recommended when no hydrogen is fed, or with any other component that is not fed.

Finally, a number should be provided to each data point, and also a number showing the chronological order of the data points, to check, if needed, whether some phenomenon is chronological.

12.3 Establishing the Correlations

A number of correlations can be built from the data set, i.e. one for each dependent variable. It is described here how to build one of these correlations. In principle, the method should be repeated in its full extent for each of the correlations to be built, though some information can be recycled when the next dependent variable is given in exactly the same set of observations as the former.

12.3.1 Inspection of the Data Set

A correlation is establishing an experimental link between a *dependent variable* Y and a number of *independent variables* X_i.
The independent variables which will be considered should be selected. Their number should be limited to those of which is known from experience that they do have an effect on the dependent variable, those which are suspected of having such an effect, and sometimes those of which should be confirmed that they do *not* have an influence. E.g. it is interesting to confirm that conversions are independent of the production rate, because that principle is greatly simplifying the reactor models. Independent variables that by their nature are highly correlated to another variable, e.g. hydrogen concentration to hydrogen to ethylene concentration ratio, should be excluded from the start. The greater the number of variables the more tedious the work will be.
All variables should be inspected, the dependent one and the independent ones: their number of data points, average value, minimum, maximum. This information will also be helpful in a later phase, to fix the range where the correlation will be valid. Only complete cases should be considered, i.e. data points where values are available for all selected variables, the dependent variable included, since only those data points will be used in the calculation.
Then the dependent variable, for those complete cases, should be inspected carefully, taking a look at the frequency histogram and the normal probability plot. If the normal probability plot shows a variable that is normally distributed, there might be little or no correlation to look for. On the other hand very odd distributions might indicate already some group of outliers.

12.3.2 Inspection of the Observations by Cluster Analysis

Calculating the matrix of observations is not very practical. Still, when facing a data set, it is necessary to know how many groups of repeated observations there are that should be considered each as one single observation, and thus to know the number of correlation coefficients that reasonably can be calculated from the data set. When facing a not so well known data set, the method of *cluster analysis* will provide an answer without calculating the matrix of observations.

Cluster analysis is grouping similar observations in groups or *clusters* which contain from 1 to n observations. The observations are considered in the p-dimensional space of the observation vectors. The starting point is that the n observations form n clusters of 1 observation each. In step 1 of the process, the two nearest observations are united in one cluster of 2 observations, and the number of clusters becomes $(n-1)$. In step two, again, the two nearest observations or clusters are united, and the number of clusters is reduced to $(n-2)$. Eventually, step $(n-1)$ will unite the last two clusters into one cluster of n observations.

The program will ask the user to select a number of options. The method should be applied to *standardised* values. The most logical method is the *centroid method*, where the position of a cluster is defined for each dimension as the arithmetical average of the components of the observations grouped in that cluster. The distance between two clusters is defined as the geometrical distance between the two positions of the clusters, or the *euclidian distance*. The last option is the *number of clusters* in which to divide the n observations.

The *agglomeration distance plot* at the number of clusters equal to 1, is offering the information sought for. It represents the process of clustering by plotting the distance between two united clusters in each step versus the number of the step. The last steps show the largest distances covered between the larger groups of similar observations. Figure 12.1 is showing the agglomeration distance plot for a demonstration data set of 40 observations, from the production of 4 different grades in the reactor, 10 observations per grade, with 7 variables. The last 3 steps 37–38–39 have united 3 clusters to 1

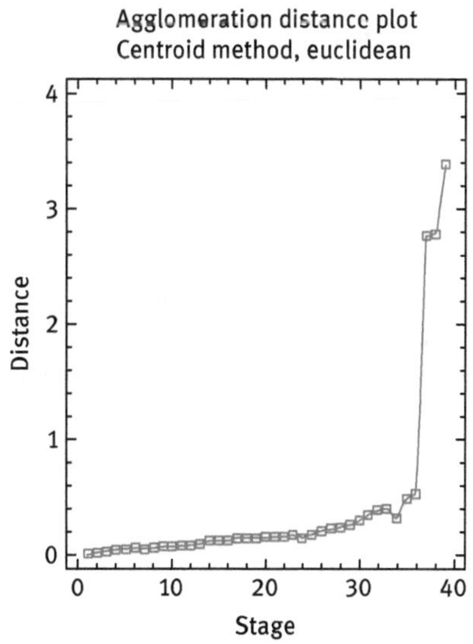

Fig. 12.1. Cluster Analysis for a data set consisting of 40 data points from 4 different reactor grades.

cluster already formed, at distances of 2.8, 2.8 and 3.4 units. It means that the number of larger clusters at considerable distances is equal to $3+1 = 4$, or the 4 grades coming to the surface. The other small distances in steps 1 to 36 are found between data points within the same grade. A correlation based on this data base should reasonably not contain more than 1 constant and 3 independent variables.

What is a considerable distance between clusters? In a space of one dimension created by one standardised variable, the maximum distance is 2 units. In two dimensions the maximum distance, between two opposite corners of the space is $2\sqrt{2}$. In p dimensions it is $2\sqrt{p}$. In the example of Figure 12.1, the maximum distance is $2\sqrt{7} = 5.3$, so a distance of 2.8 or 3.4 is significant.

In a practical data set it is recommended to view the clusters that are found at larger distances. The program will list those clusters if the optional number of clusters is set at their number. If a cluster comprises only one, or a few observations, one or more outliers may be detected already at this stage of the work and excluded, which saves time in the later work. The average values of the variables in the list of clusters will help to identify the questionable data points.

12.3.3 Inspection of the Pearson Correlation Coefficients between the Variables

In practice, the matrix of variables will not be calculated either. Instead, the Pearson correlation coefficients between all couples of variable vectors are inspected, as they are readily obtained from statistical programs. Those coefficients should be calculated for complete cases, i.e. for the set of selected independent variables X_i plus the dependent variable Y, because the Pearson correlation coefficients should be inspected for those values that will really be used when establishing the correlation.
If the total number of variables, Y together with the selected X_i, is equal to $(p+1)$, then the number of Pearson correlation coefficients to be inspected is equal to the number of combinations of two elements out of a set of $p + 1$,

$$\binom{p+1}{2} = \frac{(p+1)!}{2!(p+1-2)!} .$$

Coefficients higher than 0.80…0.85, both positive and negative, should be paid attention to. There are three possibilities when meeting such a high correlation.

High correlations between the dependent variable and an independent variable may indicate exactly the correlation that is looked for, and are perfectly alright, if they are physically acceptable.

High correlations between two independent variables may be perfectly natural, and just indicate that the selection of the independent variables should have been more restrictive. Perhaps one of the variables should be removed from the selection to avoid needless work.

High correlations between independent variables for which there is no physical explanation give something to worry about. They indicate a flaw in the data set and mean that the structure of the data set is "not orthogonal enough". A typical example of these is the negative correlation between the reactor temperature and the comonomer concentration in full scale reactor data from CrOx catalysts. There are good reasons to lower the reactor temperature when using high comonomer concentrations for low product densities, and avoid fouling of the reactor walls, hence an unwanted correlation that is often met in CrOx plant data. The presence of such a correlation in the data set should be kept in mind when establishing the correlation to give the right interpretation to the results.

If the correlation between two independent variables is really high, say > 0.95, then it is absolutely useless to look for a correlation between the dependent variable and both of the two independent variables concerned. The calculations can by no means discriminate between the effect of those two, so no effect can be assigned to either of them.

Sometimes the data set can be remedied by cutting part of the data. E.g., if there is a strong correlation between comonomer concentration and reactor temperature, it may be expedient to cut the data from one grade produced at the lowest density. The data set will, of course, be impoverished, and should be inspected all anew, as for a new data set.

Another remedy is to borrow one or more known correlation coefficients from another data set or from a part of the data set. E.g. when treating a CrOx data set, knowing that the effects of the temperature T and the activation temperature T_{act} are the same for all CrOx catalysts, these known coefficients β_1 and β_2 can be used to isolate the effect of temperature and activation temperature beforehand in the new data set. This is effected by replacing the dependent variable Y by $(Y - \beta_1 T - \beta_2 T_{act})$. Improper correlations with temperature or activation temperature in the new data set can thus be lifted. If too many unwanted high correlations occur, it may be wise to abandon the work altogether and look for a better data set.

12.3.4 Establishing the Correlation by Linear Regression

After all the preparatory work has been done, enough knowledge of the data set has been gathered to make the building of the correlation successful.

It is absolutely recommended to proceed manually, (not to allow the program to build the correlation automatically), and to apply the so-called forward selection of the variables, i.e. to start building the correlation with one variable only, and, each time after proper judgement, to add a second, a third, or more variables.

So the first step is to build p correlations between the dependent variable Y and one independent variable X_i. The technique used is multiple linear regression, and the model should comprise a constant, being of the type $\hat{Y} = \beta_0 + \beta_1 X_i$.

The following items are reviewed for each of these correlations.

1. The *plot of the observed values Y versus the predicted values* $\hat{Y}$. It gives a fast and valuable impression of the quality of the correlation. The data points should be stretched along the $Y = \hat{Y}$ line, no outliers should be at larger distances from that line. No series of data points should be visible: the data points may be present in clusters, but a series of consecutive data points, in a flight-of-stairs-like configuration is suspect.
2. The squared Pearson correlation coefficient *R-squared*, (%). In principle, the higher its value the better, though the quest for high values of R-squared may lead to taking too many variables into the correlation, and making it inaccurate, if not senseless.
3. The *p-value* for the constant and the coefficients β_i. The p-value is based on a T-distribution, and is equal to the probability that the coefficient is zero, or, in other words, not meaningful. A low p-value means that there is probably a significant relation between the dependent and independent variable. Textbooks often speak of a 95% confidence interval, in which case a two-sided p-value up to 0.0250 would be accepted. Here, when the number of data points is large enough, like for plant data of more than fifty, or rather a few hundreds of data points, a correlation should only be accepted if the p-value reads 0.0000, viz. zero in the first 4 decimal places. In case of a small number of observations, like a series of bench scale tests, larger p-values can be accepted, though with great care.
   When two correlated independent variables are introduced into the correlation, the presence of one variable will ruin the p-value of the other, and one of the two variables should be omitted. If that correlation is not explained physically, it should be noted that the effect is due to the one, or the other, or to the combination of the two variables. Such correlations are never satisfactory, and should rather not be reported. The data set should be improved instead.
4. The value of the coefficients β_i, and especially the sign of the coefficient. Is that sign physically meaningful? Is the magnitude of the coefficient in line with earlier experience? Is the sign and magnitude of the coefficient consistent during the building of the correlation? Sometimes, adding another variable to the correlation increases R-squared substantially, but alters one or more of the coefficients β_i so much, that its contribution should be suspected. The introduction of two correlated independent variables will also affect their coefficients very much.
5. The list of *unusual residuals*. The definition of a residual, proper to each data point is $(Y_j - \hat{Y}_j)$. A positive residual thus means that the observed value is above the predicted value.
   A correlation owns a certain accuracy in predicting the dependent variable for any new set of values of the independent variables. A standard error of the predicted values can be calculated, and is dependent on the position of the independent variables, i.e. will be greater near the outside of the range of the data set. The *studentised residuals* are expressed as a ratio to that standard error. Unusual resid-

uals are larger than two or three times the standard error, and may be considered as outliers.

However, a data point may be deemed an outlier by one correlation, and be normal in another. The nature and number of outliers is also an indication of the quality of a correlation, and alleged outliers should not be removed too soon.

Data points that, as the building of the correlation proceeds, consistently yield studentized residuals larger than 3.00 may be considered as outliers, and omitted from the data base. The origin of these data points should be investigated. Often they stem from unstable periods of reactor operation, e.g. from transitions or the start of a test run. At first it is better to remove only the value of the dependent variable concerned. If the data point is revealed to be an outlier for more dependent variables, it should be removed from the data set altogether. The exclusion of a group of outliers can mean a drastic improvement of the correlation, better p-values, and a more physically acceptable set of variables.

6. The *plots of residuals versus the independent variables*. Plotting the residuals in absolute values (not studentised), gives an immediate idea of how great the deviations are. Residuals should be evenly spread above and below zero, and over the whole range of the independent variable. No particular patterns must be visible in the cloud of data points. E.g., in case of a parabolic shape of the cloud, where the residuals at high and low values of the independent variable are all above or below zero, a quadratic term in the independent variable should be added ($\hat{Y} = \beta_0 + \beta_i X_i + \beta_{i1} X_i^2$). Other functions of the variables, logarithmic, exponential, are possible.

   The already mentioned series of consecutive data points, in a flight-of-stairs-like configuration, become well visible here. They may point to the influence of another variable not yet taken into consideration. Or worse, to some time related influence. E.g. low concentrations in the flash gas analysis may be very much subject to the periods between different calibrations. In such cases a plot of the independent variable versus the chronological order of the data points, should be checked to detect time dependencies. Such flaws in the data set are very difficult to repair.

The first step is concluded by the selection of one of the p correlations with one independent variable X_i as the first building block of the correlation. The correlation that has yielded the highest R-squared makes most chance to be selected, but all of the other information should not be neglected, and sound judgement should be used. The highest R-squared is not always the best choice.

The second step is to add a second independent variable to the correlation. The same procedure as described for the first step is used in this second step. So $(p - 1)$ correlations are built between the dependent variable Y, the first selected independent variable X_1 and a second independent variable $X_2 \ldots X_{p-1}$, of the type $\hat{Y} = \beta_0 + \beta_1 X_1 + \beta_i X_i$, wherefrom the second independent variable to be added to the correlation is selected.

The final correlation may be completed by the addition of a third or fourth or ... independent variable, each time repeating the same procedure. It is, however, important not to increase the number of variables beyond a reasonable number, as discussed above, and not to pursue a high R-squared at all costs. Sometimes the best results are obtained by accepting that the dependent variable is merely stochastic in the considered data range, and using its average value instead of a correlation.
The correlation might also be completed by adding the effects of interactions between the variables, i.e. new variables of the type X_iX_l. However, interactions seem to be extremely rare in reactor data.

Notations

a_{ji}	standardised value of an independent variable.
$\overline{a}_j$	observation vector.
$\underline{A}_{n,p}$	matrix of the standardised values of the independent variables.
$\underline{E}$	the unit matrix.
l_i	lowest value of one independent variable.
n	number of observations.
p	number of independent variables.
R_{il}	Pearson correlation coefficient between 2 independent variables.
T	reactor temperature, (°C).
T_{act}	chromium oxide catalyst activation temperature, (°C).
t_j	symbol representing a test or observation.
u_i	highest value of one independent variable.
$\overline{v}_i$	variable vector.
X_i	independent variable.
x_{ji}	value of an independent variable.
Y_j	dependent variable.
$\hat{Y}_j$	value of the dependent variable predicted by the correlation.
β_0	constant term in a correlation.
β_i	coefficient of an independent variable in a correlation.
θ_{jk}	angle between two vectors.

13 Scaling-Up from Bench Reactor to Loop Reactor

When a new product should be produced from a known catalyst, i.e. a catalyst for which reliable correlations have already been established from full scale plant data, the required reactor conditions can, with good accuracy, be determined from those correlations, as shown in the previous Chapters 11 and 12.
When the catalyst is not known well enough, it is common practice to predict the required reactor conditions from the results of a number of bench scale polymerisation tests. The scaling-up from bench scale to full loop reactor scale is well possible for the experienced loop reactor process engineer, and a full scale testrun can be designed on that basis. Taking an intermediate step by the use of a pilot plant for single loop production tests is expensive, time consuming, and often useless.

13.1 Bench Reactor Polymerisation Tests

The bench reactor is a small laboratory stirred tank reactor, of a volume from 2 to 10 litres. It is partially filled with diluent, thus containing two phases, a gas phase and a liquid phase. The temperature is controlled by an external cooling system. The pressure is maintained by feeding ethylene to the gas phase. A small quantity of catalyst powder is injected in the liquid, and the polymerisation starts when the proper temperature and ethylene pressure are applied. The reaction is maintained by feeding ethylene, and so controlling the pressure at a constant level. By monitoring the amount of ethylene that is fed, which is equal to the quantity consumed, the quantity of produced homopolymer is known. An example of a similar small scale homopolymerisation test was given in Section 5.5.
The effects of hydrogen and comonomers are also tested in bench scale reactors. The quantities of hydrogen and comonomer to be applied are so small that, usually, they are not fed continuously like ethylene, but are added batchwise.
The quantities of polymer collected after a polymerisation time of typically one hour, are large enough to estimate the catalyst activity, though not in absolute figures, and to determine polymer properties that do not require large samples, like melt index, density, dynamic rheology, molecular weight distribution etc. Those samples are representative, even if the particle size distribution is completely different from what it will be in full size loop production.

Bench reactor tests require a lot of experience. The greatest challenge when starting up a bench scale reactor is to get the system pure enough. Tiny amounts of poison are more effective in small systems. Therefore it is recommended to dedicate a bench reactor to one family of catalysts (Ziegler–Natta, chromium oxide (CrOx), or metallocene). Once the reactor is up and running, an experienced lab technician can produce nice results in quite a short time. To check the condition of the bench reactor it is useful to effect a standard test from time to time, with a well known catalyst in well known conditions.

13.2 Variables in Bench Reactor Polymerisation Tests

The number of *independent variables* that are involved in bench reactor polymerisations is limited, but a lot of attention should be paid to the definition of those parameters, especially when older and more recent tests are massed in one data base. Their exact meaning should be understood in detail. If tests stem from different laboratories or even different companies, though massing data from different sources in one data set is not recommended, the correct understanding of the parameters is a prerequisite.

A first group of independent variables is defining the *reactor conditions* in the bench reactor.

1. The reactor volume V_R.
2. The quantity of nitrogen in the reactor. Sometimes the initial pressure in the reactor is full vacuum at the start of the testing procedure, and no nitrogen is added. Sometimes the initial pressure is something like 1 *bar* of nitrogen, and the quantity of nitrogen present in the reactor should be calculated.
3. The type of diluent used, and its quantity, often expressed in litres at a certain temperature. This quantity can be expressed in units weight.
4. The total reactor pressure p. The reactor pressure is reached by the addition of ethylene at the start of the test, and kept constant by feeding more ethylene, which is replacing the ethylene consumed in the polymerisation.
5. The reactor temperature T.
6. The quantity of hydrogen (H_2) added to the reactor. Hydrogen is usually added batchwise at the start of the procedure. Sometimes a known volume of hydrogen is added from a separate hydrogen container at a known temperature and pressure. Sometimes the quantity of hydrogen is given as a certain hydrogen partial pressure, in the order of magnitude of 1 *bar*, established in the reactor at a specific step near the beginning of the test. The conditions at which the hydrogen is added should be noted carefully, in order to calculate the quantity of hydrogen that is present.
7. The quantity of comonomer (C_6) added to the reactor. Also the comonomer is usually added batchwise. The quantity is often expressed as a number of *millilitres* of comonomer added to the diluent.
8. The quantity of cocatalyst, if any cocatalyst is used. Its dosage is often expressed as a ratio of cocatalyst to catalyst weight, like moles per mole of Al/Ti for TEAl added to a Ziegler–Natta catalyst. This quantity of cocatalyst can be calculated,

but is a function of the quantity of catalyst present in the reactor. As explained in Section 9.2, it is preferred to use a cocatalyst concentration in the diluent y_{cocat} instead of a ratio to the catalyst weight.

This first list of independent variables is completed by a number of independent variables that, together with the reactor conditions, will determine the product properties and the catalyst activity.

1. The type and composition of the catalyst. Bench tests are often exploring the possibilities of one catalyst, but, especially with CrOx catalysts, also blends of different catalysts can be used. In such cases the composition of the catalyst blend will be a parameter. If n different catalysts are used, then $(n - 1)$ variables will be created for the weight fraction of $(n - 1)$ catalysts in the blend.
2. The quantity of catalyst or *charge*, m_{cat}, brought into the reactor at the start of the testing procedure. Its order of magnitude is 0.1 *grammes*.
3. For CrOx catalysts, the catalyst activation temperature T_{act}.
4. The reaction time, or run time, t, which here is equal to the uniform catalyst lifetime of all catalyst. The reaction is stopped at the total reaction time $t = t_1$.

The concentrations of diluent and ethylene, and the initial concentrations of hydrogen and comonomer in the bench reactor liquid are determined by the first group of independent variables. Before processing the bench reactor results, the first group of independent variables is replaced by a new group of variables, describing the composition of the bench reactor liquid.

Thereto, the quantities of added components should be carefully calculated based on the testing procedure where volumes of components and temperatures and pressures at the time of the filling of the reactor are specified. Then the concentrations of the components in the liquid should be calculated at the specified polymerisation temperature and total pressure, by calculating the vapour-liquid equilibrium (VLE) in the bench reactor volume containing these quantities of components. This is done by using the classic methods described in the chemical engineering literature, e.g. by means of a cubic equation of state.

Important results are the ethylene and hydrogen molar concentrations in the reactor liquid.

It is recommended to select total pressure and temperature in such a way that the ethylene concentration in the liquid is quite close to the ethylene concentration in the full scale loop.

A new set of independent variables is then defining the reactor conditions.

1. The type of diluent used (usually isobutane or propane).
2. The ethylene concentration in the liquid (mole%).
3. The initial (at time 0, start of the test) hydrogen to ethylene molar concentrations ratio (H_2/C_2) in the liquid.

4. The initial comonomer (1-hexene) to ethylene molar concentrations ratio (C_6/C_2) in the liquid.
5. The concentration of cocatalyst in the liquid y_{cocat}, *wtppm* of cocatalyst in the diluent, if any cocatalyst is used.

The results of a bench reactor polymerisation, or *dependent variables*, are the following.

1. The *ethylene consumption curve*, or the recorded flow of ethylene per unit time F_2 versus the run time t.
2. The *induction time*, or the time from the start of the test till the reaction starts, as seen on the ethylene consumption curve.
3. The quantity of polymer, or *yield*, m_{pol}, collected after a polymerisation time t_1 of typically one hour.
4. The catalyst productivity and activity based on m_{pol} and t_1.
5. The melt index MFI of the produced polymer powder, and of the pellets, produced on a soft lab extruder from that powder.
6. The density DENS of the produced polymer, both powder and pellets.
7. The powder particle size distribution (PSD), particle morphology, and bulk density. The PSD is completely different from what it will be in full size loop production. The bench particles are coarser due to their longer and uniform residence time. Still, the polymer particle morphology and the concentration of fines in the powder depend on the catalyst particle morphology and the catalyst activity. The powder bulk density is a related indicator. All give good information as to how suitable the reactor powder will be for loop operation.
8. The polymer properties that are related to the polymer structure, and can be measured on small quantities, like dynamic rheology, molecular weight distribution, and differential scanning calorimetry.
9. The mechanical properties of the polymer: tensional modulus, sometimes referred to as stiffness, and strength properties, which, in general, can be classified either as short term impact strength, or long term creep resistance.

Important parameters that are not predicted from bench scale polymerisations are the drops or increases in melt index and density that are observed when extruding the reactor powder on full scale. Full scale extrusion will often modify the product. The changes in melt index and density encountered there, can only be learnt from experience. They can surprise the test run engineer and necessitate adjustments in the reactor conditions. Good correlations will facilitate the corrective actions.
The extrusion of bench reactor powder, however, should be soft enough, homogenising the product while not changing its properties.

13.3 Catalyst Activity

The catalyst activity is an important result from the bench test, and deserves some special attention.

In bench tests it is common to define the *catalyst productivity*, $\mathbb{P}$, as the total weight of polymer, m_{pol}, produced at time t_1 when the reaction is stopped, per unit catalyst weight, m_{cat}, in the reactor,

$$\mathbb{P} = \left(\frac{m_{\text{pol}}}{m_{\text{cat}}}\right)_{t=t_1},$$

and the *catalyst activity*, $\mathbb{A}$, as the time average over the reaction time t_1 of the weight of polymer produced per unit catalyst weight, m_{cat}, in the reactor,

$$\mathbb{A} = \left(\frac{m_{\text{pol}}}{m_{\text{cat}} t_1}\right)_{t=t_1} = \frac{\mathbb{P}}{t_1}.$$

These definitions are flawed since the values depend on the total reaction time t_1.

Instead of assigning a single value to the catalyst activity, it is more informative to look at the whole activity curve recorded during the test.

The bench reactor is a batch reactor with respect to the catalyst, and, as explained in Section 5.5, the catalyst reactor activity $\mathbb{A}(t, Rx)$ is identical to the catalyst activity $\overline{\mathcal{A}}(t, Rx)$, due to the uniform residence time. This activity is equal to the ethylene feed flow, F_2, per unit catalyst weight,

$$\mathbb{A}(t, Rx) = \overline{\mathcal{A}}(t, Rx) = \frac{F_2(t)}{m_{\text{cat}}}.$$

From what has been shown on catalyst activity in the previous chapters, it is expected that this activity curve will be of the form,

$$\overline{\mathcal{A}}(t, Rx) = \frac{F_2(t)}{m_{\text{cat}}} = b\mathrm{e}^{-at}.$$

This is most often not the case ! Three types of activity curves, commonly obtained from bench polymerisation tests, are shown in Figure 13.1. Type 1 is the expected exponentially declining curve, apart from some start-up phenomena. Type 2, very common with CrOx catalysts, is increasing within times up to the usual residence times in commercial loop reactors. Type 3 is increasing also and then stabilising for a longer time before a slow decline will start.

It is clear that the types 2 and 3 cannot be valid in full scale loop reactors.

General experience and the full scale tests described in Chapter 8 and illustrated in Figures 8.1 to 8.4 show that the catalyst productivities in the full scale loop reactor are increasing with residence time at a rate less than linear. So the catalyst activity curve can only decline with residence time, and not increase.

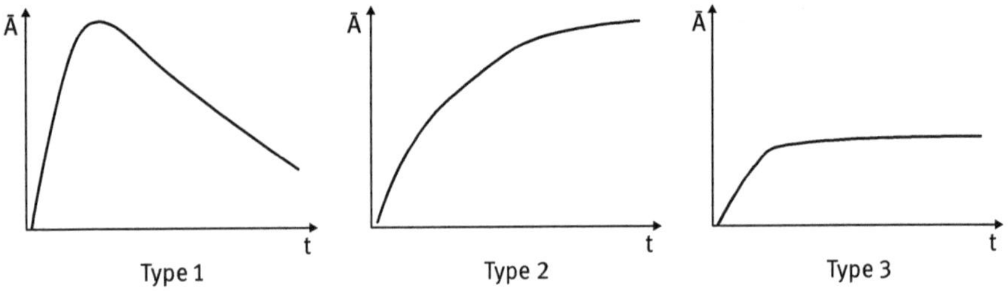

Fig. 13.1. Three types of activity curves from bench polymerisation tests.

Catalyst activity curves from bench tests at different temperatures, or at different ethylene partial pressures, all other conditions kept constant, may show the three types of curves, depending on the selected temperature or ethylene partial pressure[1].
These large deviations can be understood when realising that the overall polymerisation kinetics are not only determined by the polymerisation reaction step itself, but also, and probably predominantly, by the diffusion steps: diffusion of monomer from gas- to liquid phase; in the liquid phase; from liquid to polymer particle; in the polymer phase to the active site. Also the solubility of the monomer in the polymer, which influences its diffusion in that medium, is temperature dependent. As a consequence it can never be guaranteed that the kinetics in the bench reactor are identical to the kinetics in the loop.
In spite of this, sometimes complex models are constructed, making use of large numbers of kinetic constants to match the bench reactor activity curves, with the goal of building reactor models to control the full scale reactor. The usefulness of these models is doubtful. The simple and direct empirical approach to determine the catalyst activities from full scale tests, as described in Chapter 8, is much more practical, and perhaps the only way to obtain valuable results.

The catalyst activity is thus an important result from the bench test, but should not be treated as just another dependent variable. It is not the catalyst activity that will be obtained in the loop, and it does not make sense to optimise it versus the reactor conditions. It is rather a boundary condition to the problem of selecting the reactor conditions where the best polymer is made: by comparing bench scale activities in different tests, it should simply be checked whether the catalyst activity is acceptable for commercial production.

1 See e.g. for the temperature influence on catalyst activity in a gas phase bench polymerisation of butadiene, *Zöllner and Reichert, Chemie Ingenieurtechnik, volume 72, p. 399, 2000.*

13.4 Data Set and Correlations

Bench scale tests are effected to solve the problem of producing the optimal product from a certain catalyst. What is optimal depends on the desired polymer properties. Often not a single property will be optimised but a compromise between different properties will be sought, as e.g. between short term impact strength and long term creep resistance.

The optimal polymer from a certain catalyst will always correspond with a polymer of a certain melt index MFI and density DENS from that catalyst. Scaling-up from bench reactor to full scale loop means to translate the bench reactor conditions where the optimal product is produced into full scale loop reactor conditions that will yield the same product.

The first step in the scaling-up is to compose a well-balanced data set.

When collecting data from laboratory polymerisations one has the great advantage that the tests can be designed in number and in choice of parameters. On the other hand, the number of tests is limited due to the work and time involved to get a lab scale reactor running and effect the tests. It is absolutely requisite to follow a factorial design of experiments scheme to specify the tests, and obtain variable vectors that are orthogonal (see Chapter 12). It is also preferable to repeat the tests.

If the test scheme cannot be influenced any more, e.g. when using old data, the best evaluation of the scheme is to check how well it fits into a factorial design. Too often the investigator is hoping to produce the right product in his lab and repeating a certain test at the supposed target a number of times at the expense of other tests. It is much better to explore the process window by a well balanced design, which allows to establish correlations, and to find an optimum, which can perhaps be confirmed by a second small series of additional tests near the optimum.

Very often the researchers will set up a new design of experiments to investigate a new problem, and try and resolve the problem only from that information. They seem to forget all of the other tests effected in the past. However, if older bench scale results are available on the same catalyst system, it would be a true waste of resources not to use those precious results. If those tests were valuable once, there is no reason to doubt their validity later. Combining the new tests with the old ones will provide a more extended data base and allow a better judgement when drawing conclusions. Still, the compatibility of the test series should be checked. It is possible that the definition of some parameters has changed.

The second step is to build correlations between each dependent variable and the independent variables.

The dependent variables are melt index MFI, density DENS, and the polymer properties that are of interest. It is strongly recommended to use the values measured on the pellets, and not on the powder. Those pellets are more homogeneous and thus more representative for the product, since they have been extruded on a soft lab extruder. These dependent variables are correlated with the following independent variables, i.e. with the most important bench reactor conditions: the composition of the catalyst; the reactor temperature; the (CrOx) catalyst activation temperature; the ethylene concentration; the concentration ratios of hydrogen and comonomer to ethylene in the bench reactor liquid; the cocatalyst concentration.

Even if it is strongly recommended to use a factorial design of experiments scheme to structure the data set, it is not at all recommended to use the traditional concepts from factorial experiments to establish the correlations. The calculation of *effects* and *interactions* as practiced in the area of factorial experiments, is very simple but shows a great lack of flexibility. Merging two series of experiments, or meeting some variation in the independent variables is causing difficulties that can hardly be overcome. So factorial design is a good help to set up valuable data sets, but to calculate the correlations, linear regression methods, which easily handle all data sets, should be used. The guidelines given in Chapter 12 are fully applicable here. Bench correlations are of the same form as loop correlations. Interactions between variables have never been met by the author.

From those correlations the optimal catalyst composition and product melt index and density are selected, together with the bench reactor conditions that will yield this optimal product. These conditions should now be transposed to the full scale loop reactor.

13.5 Translating Bench Scale Variables into Loop Reactor Conditions

How to find the corresponding value in the loop reactor for every independent variable in the bench reactor, is discussed in this section. The idea is not only to translate one optimal set of bench conditions into one set of loop conditions, but preferably to transform bench reactor correlations into loop correlations. The latter are helpful to steer testruns towards the target product.

13.5.1 Type and Composition of the Catalyst

The product obtained with a certain type and composition of the catalyst will be reproduced in the loop.

13.5.2 Ethylene Concentration

The ethylene concentration in the loop reactor liquid is equal to the concentration in the bench reactor liquid.
The ethylene concentration is often not explicitly mentioned when bench results are

reported, because the independent variables, especially the total pressure, are selected in such a way that the ethylene concentration in the bench liquid is quite close to the most typical ethylene concentration in the full scale loop.

13.5.3 Reactor Temperature

It is expected that temperature measurements in different places in the world will not show great deviations. However, experience has taught that large temperature differences are possible between bench reactors and full scale loop reactors producing the same product quality. The temperature is not that critical when running Ziegler–Natta and metallocene catalysts, but with CrOx catalysts, a shift in the required reactor temperature can ruin a whole testrun. Applying the same temperature to the loop as in the bench reactor may there cause great disappointments. This has probably to do, not with erroneous thermometers, but with different catalyst poison levels in the systems. The purity of a large scale process system can never be equalled in small laboratory setups. So, the only way to learn how the bench temperature should be transferred to the loop, is by comparing bench results with full scale loop data, and building up experience.

13.5.4 Catalyst Activation Temperature

For CrOx catalysts, the catalyst activation temperature T_{act} is an important parameter. Normally, the small scale activations of catalysts to be used in the bench reactor are well representative for the full scale activations, and the same activation temperature is applied there.

13.5.5 Hydrogen Concentration

The initial hydrogen to ethylene molar concentrations ratio (H_2/C_2) in the bench reactor liquid can be applied directly to the loop conditions without any transformation. This seems acceptable for CrOx catalysts, where the hydrogen conversion is very low, and where most probably not much of the initial quantity of hydrogen is consumed. But also for Ziegler–Natta and chromocene catalysts this principle holds.

The hydrogen incorporated in the polymer, x_H, is only a small fraction of the polymer, some 100 *wtppm*, and often less. The consumption of hydrogen during the bench test is also small, order of magnitude of 0.01 *moles* of hydrogen. When calculating vapour-liquid equilibria for a few practical cases, where bench reactor tests are effected with hydrogen, it becomes clear that this consumption constitutes only some 10% of the initial hydrogen quantity. It will also be noted that although the vapour phase is small compared to the liquid phase, it is acting as a buffer and delivering the consumed hydrogen. The hydrogen concentrations in the liquid stay more or less constant, while the consumed hydrogen is drawn from the shrinking vapour phase, due to the vapour liquid equilibria that are restored. The important consequence is that the (H_2/C_2) molar ratio in the bench reactor liquid is practically constant during the polymerisation, and can be transferred directly to the loop.

A special case is formed by the metallocene catalysts where the hydrogen conversion is 100%. There the incorporated hydrogen is known and can be linked to the melt index of the product.

13.5.6 Comonomer Concentration

The comonomer conversion and the comonomer incorporated in the polymer cannot be determined from bench reactor data, neither are they known for an unknown catalyst in the loop. It is not easy to translate a quantity expressed as a number of *millilitres* of comonomer, added batchwise to the diluent in the bench reactor, to a comonomer concentration in a continuous loop reactor. Still, the decreasing comonomer concentration in the bench reactor, resulting from a single initial dosage, can be called *equivalent* to a constant comonomer concentration in the loop reactor *when in both cases the same polymer density is produced.*

To find the comonomer-, say 1-hexene concentration, in the loop, that is equivalent to the 1-hexene dosage in the bench reactor, a density correlation for a specific catalyst from the loop is compared to its corresponding density correlation in the bench. The constant full scale 1-hexene concentration is expressed by the 1-hexene to ethylene concentration ratio (C_6/C_2) in the flash gas.

The density correlation for the loop reactor, see Chapter 9, Section 9.5.2, is of the form,

$$\text{DENS} = \beta_0 - \beta_1 (x_6)^b + \beta_2 \log(\text{MFI}) .$$

An equivalent correlation in terms of (C_6/C_2) can be established from the same loop data,

$$\text{DENS} = \beta_{0,\text{loop}} - \beta_{1,\text{loop}} (C_6/C_2)^b + \beta_2 \log(\text{MFI}) . \quad (13.1)$$

This correlation (13.1) is less precise than the former one, because it does not take into account the variations in 1-hexene conversion, that are mainly due to variations in the ethylene concentration. However, testruns with new catalyst systems are normally run at mid range ethylene concentrations, avoiding extreme values. Also, the level of precision expected in this scale-up work is below what is required to optimise a commercial running production.

Let a density correlation be known for the same catalyst system from a series of tests in the bench reactor, and in terms of the added quantity of comonomer, (C_6). It should be of the form,

$$\text{DENS} = \beta_{0,\text{bench}} - \beta_{1,\text{bench}} (C_6)^b + \beta_2 \log(\text{MFI}) . \quad (13.2)$$

Note that the coefficient β_2, for the effect of the MFI on density, is taken as a constant within one family of catalysts.
When both 1-hexene concentrations are equivalent, both densities should be equal at the same value of the melt index. By equating Equations 13.1 and 13.2, the relation between the 1-hexene to ethylene concentration ratio (C_6/C_2) in the loop, and the added quantity of comonomer to the bench, (C_6), is obtained,

$$(C_6/C_2) = \left(\frac{\beta_{0,\text{loop}} - \beta_{0,\text{bench}} + \beta_{1,\text{bench}}(C_6)^b}{\beta_{1,\text{loop}}} \right)^{1/b} . \tag{13.3}$$

Equation (13.3) provides the translation from comonomer dosage in the bench reactor to loop reactor comonomer concentration.

An alternative way to estimate the 1-hexene concentration, when not enough information is available to follow the above method, is to estimate the ethylene and 1-hexene conversions in the loop by assuming that these conversions will be the same as for a known catalyst from the same catalyst family, and that the unknown catalyst will also obey the loop reactor density correlation for that catalyst family, Chapter 9, Section 9.5.2,

$$\text{DENS} = \beta_0 - \beta_1(x_6)^b + \beta_2 \log(\text{MFI}) .$$

The incorporated 1-hexene, x_6, is obtained from this correlation, and the required 1-hexene to ethylene ratio in the flash gas, (C_6/C_2), is calculated from equation (11.2),

$$(C_6/C_2) = \frac{y_6}{y_2} = \frac{\xi_2}{\xi_6}\frac{1-\xi_6}{1-\xi_2}\frac{x_6}{1-x_6-x_\text{H}}\frac{28.05}{84.16} \approx \frac{\xi_2}{\xi_6}\frac{1-\xi_6}{1-\xi_2}\frac{x_6}{1-x_6}\frac{28.05}{84.16} .$$

No information from bench scale tests is involved here, but the risk is real that the true conversions and densities in the full scale testrun will be quite different. The conversions should then be checked during the testrun, which is always useful, and corrections made, based on those values. These corrections will necessarily take an extra step in the testrun.

13.5.7 Dosage of the Cocatalyst

When making Ziegler–Natta polymers in the bench reactor, high dosages of cocatalyst (TEAl) are used, mainly to fight the higher poison levels in the system. Part of this dosage may be consumed even, upon reacting with these poisons, especially with moisture. So, instead of copying the cocatalyst concentrations from the bench conditions, the usual cocatalyst levels should be maintained in the loop. These levels have been optimised to obtain the highest catalyst activity. The catalyst will not be active enough at too low levels, and beyond a certain optimum, its activity will drop again due to a poisoning effect from too high dosages.
The cocatalyst concentration does influence the reactor conditions required to obtain a certain melt index and density, through the conversions of ethylene and hydrogen.

But the polymer, at the right melt index and density, will be very much the same, regardless of the applied cocatalyst level. Adjustments to the reactor conditions in the loop will have to be made anyway, during the first full scale run, and will implicitly take account of the used cocatalyst level.

It is different when cocatalysts, like TEB, are used together with CrOx catalysts. The effect of the cocatalyst on the catalyst activity, the melt index and the density, can be large and is not predictable. It can even work in both ways, increasing or decreasing. There it is necessary to translate the usual *moles* per *mole* of B/Cr to a cocatalyst concentration in the diluent y_{TEB}, and to apply the same concentration in the loop.

13.5.8 Comparing Bench Results to Full Scale Loop Data

Whenever some full scale loop information on the studied catalyst is available, albeit only one data point, it should be checked whether this data is matching the bench results. It is very useful to plot curves of the bench correlations, and to add points to them, if available, from loop data. A good match will be reassuring, whilst a mismatch may indicate some mistake.

Besides data points, it is also interesting to compare trends between bench and loop. The coefficients of the same variables should be more or less equal for both. The melt index e.g., should show the same slope as a function of reactor or activation temperature in bench and loop environments.

With CrOx catalysts, where the effects of the reactor parameters are equal for all catalysts, correlation coefficients well known in the loop can be used when building the bench correlations and be a valuable supplement to the information available from the bench.

13.6 Molecular Weight Distributions from Bench Scale Tests

The polymer samples from bench scale tests are representative for what will be obtained in the loop reactor. Their properties will be reproduced. Unfortunately, researchers will, in this expectation, often look for the optimal product sample, and neglect the investigation of a broader process window. Still, samples from a wider range of reactor conditions will provide much better data to support a full scale run, and yield a more profound knowledge of the polymer properties. This is illustrated here, by the example of the molecular weight distribution.

The molecular weight distribution (MWD) is an important characteristic of the polymer. It is measured by gel permeation chromatography (GPC), of which the result is most often presented graphically, by a continuous curve, viz. the distribution curve $\mathrm{d}W/\mathrm{d}M$ plotted versus M, the molecular weight of the polymer, and where $\mathrm{d}W$ is the fraction of the polymer that owns a molecular weight between M and $(M + \mathrm{d}M)$.

The molecular weight M is presented on a logarithmic scale $\log_{10} M$.
However valuable this information may be, it does not tell everything.
The MWD curve is quite accurate in the low molecular weight (LMW) range, but lacks precision in the very high molecular weight (HMW) end, which is narrowed by the use of the logarithmic scale, which actually "hides" a long tail of fractions of an extremely long chain length. These are detected more easily by the elastic behaviour of the polymer in dynamic rheology tests.
The MWD curve does not describe the distribution of the comonomer either. The short side chains added to the polymer by the comonomer, the so called short chain branching (SCB), are simply not detected by the GPC analysis.

The MWD of a high density polyethylene or a polypropylene product can be simplified and represented by a discrete function consisting of some nine or ten data points. For products produced from the same catalyst, the discrete GPC curves can be correlated with log(MFI), the logarithm of the melt index. This means that, based on a limited number of bench reactor samples, the whole family of GPC curves of a type of product is known.
Let,

- GPC_{min} and GPC_{max} be the $\log_{10} M$ values for the onset and endpoint of the MWD curve, where the value of the distribution curve is zero;
- GPC_{xxx} be the values of the MWD curve at $\log_{10} M = x.xx$.

Each of these discrete values can be represented by a linear correlation with log(MFI). An example of 5 MWD curves measured on 5 bench reactor samples from one CrOx catalyst, is shown in Fig 13.2. This rather confused bunch of curves becomes much clearer when presented as a set of linear correlations to log(MFI), as has been done for 7 GPC_{xxx} values in Fig 13.3.
Here, as in many polyethylene products, there is no correlation found for GPC_{500}, but its value is rather a constant. GPC_{500} is often a pivot point in a family of MWD curves, where all curves intersect.
The positive slope for GPC_{450} and the negative slope for GPC_{550} show how an increase in melt index results in a shift of material from the area around $M = 10^{5.5}$ to $M = 10^{4.5}$.
Certain strength properties of the polymer can be correlated with certain GPC_{xxx} values, but not exclusively to those, as the MWD does not describe the polymer completely. The crystallinity is another important aspect.

The MWD curve of the optimal product can be obtained by interpolation at the target value of log(MFI). A smooth MWD curve, going through the discrete points, can easily be drawn, to reconstruct and visualize the MWD in the familiar graphical form. When the full scale product will be available, it will be seen that its discrete MWD curve does fit the correlations. This will be the case for a homopolymer, and just as well for a copolymer, since short chain branching is not detected in the GPC analysis.

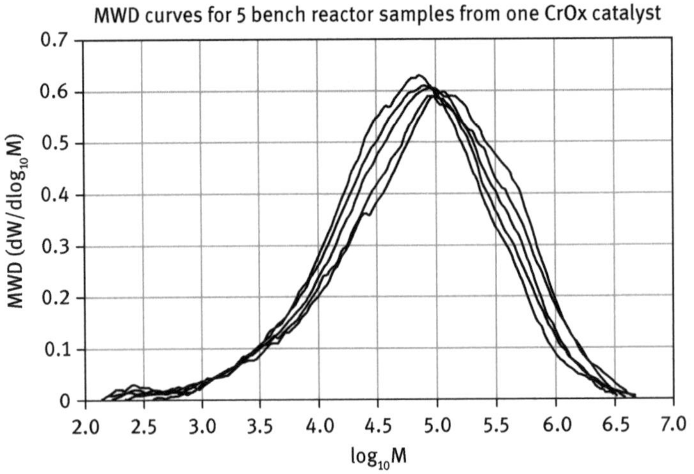

Fig. 13.2. MWD curves by GPC measured on 5 bench reactor samples from one CrOx catalyst.

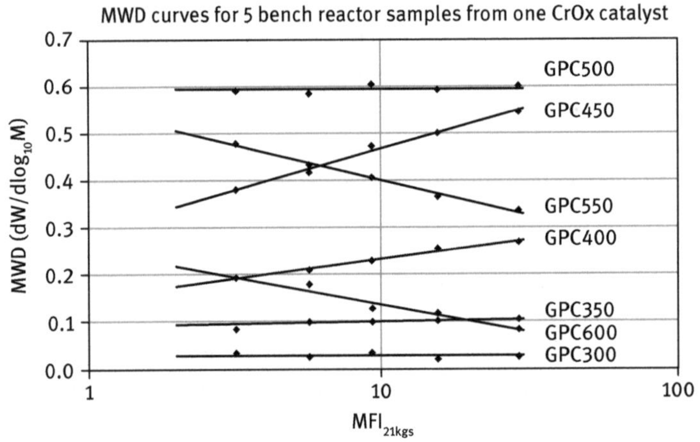

Fig. 13.3. Linear correlations of the GPC_{xxx} values with log(MFI) for the MWD curves of 5 bench reactor samples, shown in Figure 13.2.

Notations

The reactants and chemical substances are indicated by a subscript:

2	ethylene
6	1-hexene
H	hydrogen
cat	catalyst
cocat	cocatalyst

$\mathbb{A}$	catalyst reactor activity: quantity of polymer produced by all catalyst present in the reactor per unit time, per *kg* catalyst, (*kg* polymer/*kg* catalyst, *s*).
$\overline{\mathcal{A}}(t, Rx)$	catalyst activity: quantity of polymer produced by one *kg* of catalyst per unit time at a certain lifetime *t* in the bench reactor, (*kg* polymer/*kg* catalyst, *s*).
(C_6)	quantity of comonomer added batchwise to the bench scale reactor, (*mole* 1-hexene).
(C_6/C_2)	1-hexene to ethylene molar concentration ratio in the (loop or bench) reactor liquid, (*mole* 1-hexene/*mole* ethylene).
DENS	density of the polymer measured in the lab at room temperature, (kg, m^{-3}).
F_2	ethylene feed to the bench reactor, (kg, s^{-1}).
GPC_{max}	$\log_{10} M$ value for the endpoint of the MWD curve, where the value of the distribution curve is zero.
GPC_{min}	$\log_{10} M$ value for the onset of the MWD curve, where the value of the distribution curve is zero.
GPC_{xxx}	value of the MWD curve at $M = 10^{x.xx}$, ($kg, kg^{-1}, g^{-1}, mole$).
(H_2)	quantity of hydrogen added batchwise to the bench scale reactor, (*mole* hydrogen).
(H_2/C_2)	hydrogen to ethylene molar concentration ratio in the (loop or bench) reactor liquid, (*mole* hydrogen/*mole* ethylene).
M	molecular weight of the polymer, ($g, mole^{-1}$).
m_{cat}	mass of catalyst present in the bench reactor, (*kg*).
m_{pol}	mass of polymer present in the bench reactor, collected after the polymerisation time t_1, (*kg*).
MFI	melt index of the polymer, (dg, min^{-1}).
p	reactor pressure, (*Pa*).
$\mathbb{P}$	catalyst productivity: total quantity of polymer produced by one *kg* of catalyst at the end of the reaction time t_1, (*kg* polymer/*kg* catalyst).
Rx	symbol representing all of the reactor conditions, including temperature, concentrations, flows, and densities.
t	catalyst lifetime: the time interval between the moment when the bench reaction is started, and any moment at which the reactor system is considered, (*s*).
t_1	(uniform) catalyst lifetime at which the reaction of a bench test is stopped, (*s*).
T	reactor temperature, (°C).
T_{act}	chromium oxide catalyst activation temperature, (°C).
V_{R}	reactor volume, (m^3).

W	fraction of the polymer that owns a molecular weight between M and $(M + \Delta M)$, (kg, kg^{-1}).
x_6	incorporated 1-hexene in the polymer, (*kg* 1-hexene/*kg* polymer).
x_H	incorporated hydrogen in the polymer, (*kg* hydrogen/*kg* polymer).
y_{cocat}	cocatalyst concentration, (*kg* cocatalyst/*kg* reactor liquid).
y_{TEB}	TEB cocatalyst concentration, (*kg* TEB/*kg* reactor liquid).
β_0	constant term in a correlation.
β_i	coefficient of an independent variable in a correlation.

14 The Operation of Two Loop Reactors in Series

In this last chapter the operation of two loop reactors in series is discussed. In such an operation the polymer produced in the first reactor is fed to a second reactor, where the catalyst contained in this polymer will resume its activity, if the proper reactor conditions are offered. The reason for adopting that kind of operation is often to obtain a so-called *bimodal product*. When the reactor conditions in the second reactor are different from those in the first, the polymer produced in total will show a broad molecular weight distribution (MWD), being a combination of the two different MWD's produced in the two reactors. Such a broad product may show certain advantages in properties.

However, the operation of two reactors in series brings also clear disadvantages. The residence time in the second reactor is limited by the product flow from the first reactor, and so the second reactor is less productive, or must be more largely sized. Also, the idea that, in a bimodal production, the polymers of the two different MWD's are most intimately mixed, because both are formed on each catalyst particle, is rather illusive. Due to the nature of the residence time distributions in loop reactors, most particles will carry a polymer that is predominantly formed in the one or the other reactor.

The following assumptions are made here.

1. Steady state in both reactors. The polymer flow fed to the second reactor is equal to the production rate in the first reactor.
2. The 1/1 hypothesis. The catalyst feed rate to the first reactor is transferred by the product flow to the second reactor. So the catalyst feed rates to both reactors are identical.
3. Perfect mixing of the catalyst in both reactors.

Due to these assumptions, especially perfect mixing, all discussions that follow are also valid for those bimodal processes where a gas-phase (fluidized bed) reactor is combined with a loop reactor, because perfect mixing is even more obviously valid for a gas-phase reactor. This means that the disadvantages mentioned above are also true for such processes, but, on top of that, such hybrid systems will be very complex to operate.

14.1 Mass of Polymer Present in the Second Reactor

According to the 1/1 hypothesis, the *total mass of polymer in the second reactor*, $m_{\text{pol},2}$, is equal to the total mass of polymer that has been produced by all catalyst particles present in the second reactor. It must be noted that this total mass of polymer does include the polymer that has been produced in both the first and second reactors.

In Section 7.5 it is shown that, under the above assumptions, the mass of polymer present in the reactor is equal to the mass of catalyst in the reactor times the catalyst productivity. When applying this to the second reactor, the catalyst productivity should be replaced by the sum of the productivities in both reactors,

$$m_{\mathrm{pol},2} = m_{\mathrm{cat},2}(\mathbb{P}_1 + \mathbb{P}_2)\,, \tag{14.1}$$

where $m_{\mathrm{cat},2}$ represents the total mass of catalyst present in the second reactor, and $\mathbb{P}_1$ and $\mathbb{P}_2$ the catalyst reactor productivities in the first and the second reactor respectively.

14.2 Catalyst Residence Time Distribution in the Second Loop Reactor

The calculation of the residence time distribution in the second reactor is analogous to the one for the first reactor, Section 7.6.
Suppose a small fraction of the *catalyst feed flow*, F_{cat}, marked by some tracer property, and injected at time $\theta_2 = 0$ in the second reactor, is followed on its path through that reactor. Let its quantity be $q_0 = F_{\mathrm{cat}}\mathrm{d}\theta_2$.
θ_2 is the *catalyst residence time in the second reactor.*
Let the *quantity of the tracer catalyst present in the second reactor* be q_2.
The concentration of the tracer catalyst in the polymer in that reactor is, making use of (14.1), equal to,

$$\frac{q_2}{m_{\mathrm{pol},2}} = \frac{q_2}{m_{\mathrm{cat},2}(\mathbb{P}_1 + \mathbb{P}_2)}\,.$$

The stream of polymer leaving the second reactor is $F_{\mathrm{cat}}(\mathbb{P}_1 + \mathbb{P}_2)$.
A mass balance for the tracer catalyst, taking into account that all catalyst is present in the solid (polymer) phase, is,

$$-\mathrm{d}q_2 = \frac{q_2}{m_{\mathrm{cat},2}(\mathbb{P}_1 + \mathbb{P}_2)} F_{\mathrm{cat}}(\mathbb{P}_1 + \mathbb{P}_2)\mathrm{d}\theta_2\,,$$

or,

$$\frac{\mathrm{d}q_2}{q_2} = -\frac{F_{\mathrm{cat}}}{m_{\mathrm{cat},2}}\mathrm{d}\theta_2\,,$$

which is solved as,

$$q_2 = q_0 \exp\left(\frac{-F_{\mathrm{cat}}}{m_{\mathrm{cat},2}}\theta_2\right)\,.$$

The *catalyst residence time distribution in the second reactor*, $E_{2,2}(\theta_2)$, is a function of θ_2 such that $E_{2,2}(\theta_2)\mathrm{d}\theta_2$ is equal to the weight fraction of the catalyst in the second reactor with a residence time between θ_2 and $(\theta_2 + \mathrm{d}\theta_2)$,

$$E_{2,2}(\theta_2)\mathrm{d}\theta_2 = \frac{-\mathrm{d}q_2}{q_0} = \frac{1}{q_0} q_2 \frac{F_{\mathrm{cat}}}{m_{\mathrm{cat},2}}\mathrm{d}\theta_2\,.$$

Hence,

$$E_{2,2}(\theta_2) = \frac{F_{\text{cat}}}{m_{\text{cat},2}} \exp\left(\frac{-F_{\text{cat}}}{m_{\text{cat},2}}\theta_2\right) . \tag{14.2}$$

The *average catalyst residence time in the second reactor*, τ_2, is calculated as,

$$\tau_2 = \int_{\theta_2=0}^{\infty} \theta_2 E_{2,2}(\theta_2) d\theta_2 = \frac{m_{\text{cat},2}}{F_{\text{cat}}} .$$

The average catalyst residence time τ_2 can also be expressed in terms of the polymer as,

$$\tau_2 = \frac{m_{\text{cat},2}}{F_{\text{cat}}} = \frac{m_{\text{cat},2}(\mathbb{P}_1 + \mathbb{P}_2)}{F_{\text{cat}}(\mathbb{P}_1 + \mathbb{P}_2)} = \frac{m_{\text{pol},2}}{P_1 + P_2} ,$$

where P_1 and P_2 are the *polymer production rates* in the first and second reactor respectively.

The *catalyst residence time distribution in the second reactor* has thus been found as,

$$E_{2,2}(\theta_2) = \frac{1}{\tau_2} \exp\left(\frac{-\theta_2}{\tau_2}\right) ;$$
$$\tau_2 = \frac{m_{\text{cat},2}}{F_{\text{cat}}} = \frac{m_{\text{pol},2}}{P_1 + P_2} . \tag{14.3}$$

The expression for the average residence time in the second reactor shows how it is limited by the production of the first reactor passing through the second loop and demanding its share of the reactor volume.

Example

To illustrate how important this reduction in residence time is, two reactors in series are considered, of identical size, at identical conditions, and both at the maximum solids concentration, thus containing the same mass of polymer. A constant catalyst activity profile $\overline{A}$ is assumed in both reactors,

$$\tau_1 = \frac{m_{pol,1}}{P_1} ; \quad P_1 = F_{\text{cat}}\tau_1\overline{A} ;$$
$$\tau_2 = \frac{m_{\text{pol},2}}{P_1 + P_2} ; \quad P_2 = F_{\text{cat}}\tau_2\overline{A} .$$

It follows that,

$$\frac{P_2}{P_1} = \frac{\tau_2}{\tau_1} = \frac{P_1}{P_1 + P_2} ,$$

whence,

$$\frac{P_2}{P_1} = 0.62 .$$

The residence time in the second reactor, and thus its production capacity, is limited to a little more than 60% of that in the first reactor.

On the other hand, there is a gain in catalyst productivity,

$$\mathbb{P}_1 + \mathbb{P}_2 = \frac{P_1 + P_2}{F_{\text{cat}}} = \frac{1.62 P_1}{F_{\text{cat}}} ,$$

compared to operating the two identical reactors in parallel, where the production would be equal to $2P_1$, and where,

$$\mathbb{P}_1 = \mathbb{P}_2 = \frac{P_1}{F_{cat}} ,$$

would be the productivity. □

14.3 Total Catalyst Residence Time Distribution in Two Loop Reactors in Series

A distribution can be calculated for the total residence time in two reactors in series, irrespective of the time being spent in the first or the second reactor. This function only makes sense when identical reactor conditions prevail in both reactors. In such case the two reactors in series should produce an identical product, though it is difficult to realise absolutely identical conditions in two separate reactors. An alternative is to use a larger single reactor or to operate the two reactors in parallel. As illustrated in the example in the previous Section 14.2, operation in series will yield a higher catalyst productivity at a reduced production capacity, whilst operation in parallel yields a lower catalyst productivity at a higher capacity.

The calculation of the total residence time distribution in two reactors in series is well known from chemical reactor theory.

The *total residence time*, θ, is the sum of the two residence times in the separate reactors,

$$\theta = \theta_1 + \theta_2 .$$

The fraction of the catalyst with a total residence time θ, which is the sum of residence times θ_1 and θ_2, is,

$$E_{2,1}(\theta_1)\mathrm{d}\theta_1 E_{2,2}(\theta_2)\mathrm{d}\theta_2 ,$$

or,

$$E_{2,1}(\theta_1)\mathrm{d}\theta_1 E_{2,2}(\theta - \theta_1)\mathrm{d}\theta .$$

The *total catalyst residence time distribution in two reactors in series*, $E_{2,1+2}(\theta)$, is a function of the total catalyst residence time θ in the two reactors such that $E_{2,1+2}(\theta)\mathrm{d}\theta$ is equal to the weight fraction of the catalyst leaving the second reactor with a total residence time in the two reactors between θ and $(\theta + \mathrm{d}\theta)$. This fraction is the sum of all catalyst fractions with any residence time θ_1 and the appropriate θ_2. The value of θ_1 cannot be larger than θ,

$$E_{2,1+2}(\theta)\mathrm{d}\theta = \int_{\theta_1=0}^{\theta} E_{2,1}(\theta_1)\mathrm{d}\theta_1 E_{2,2}(\theta - \theta_1)\mathrm{d}\theta .$$

Such an integral is known as the convolution integral of $E_{2,1}(\theta_1)$ and $E_{2,2}(\theta_2)$ over the interval from 0 to θ.

$$E_{2,1+2}(\theta)\mathrm{d}\theta = \int\limits_{\theta_1=0}^{\theta} \frac{1}{\tau_1} \exp\left(\frac{-\theta_1}{\tau_1}\right)\mathrm{d}\theta_1 \frac{1}{\tau_2} \exp\left(\frac{-(\theta-\theta_1)}{\tau_2}\right)\mathrm{d}\theta \,.$$

Whence is obtained,

$$E_{2,1+2}(\theta) = \frac{1}{\tau_1 - \tau_2}\left[\exp\left(\frac{-\theta}{\tau_1}\right) - \exp\left(\frac{-\theta}{\tau_2}\right)\right] . \tag{14.4}$$

The *average total residence time*, τ_{1+2}, is, not surprisingly,

$$\tau_{1+2} = \int\limits_{\theta=0}^{\infty} \theta E_{2,1+2}(\theta)\mathrm{d}\theta = \tau_1 + \tau_2 \,.$$

In case that $\tau_1 = \tau_2 = \tau$, the expression (14.4) for the total catalyst residence time distribution is not defined. $E_{2,1+2}(\theta)$ is obtained by repeating the above calculation for that particular case,

$$[E_{2,1+2}(\theta)]_{\tau_1=\tau_2=\tau} = \frac{1}{\tau^2}\theta \exp\left(\frac{-\theta}{\tau}\right) . \tag{14.5}$$

14.4 Assumed Catalyst Productivity in a Second Loop Reactor in Series

Knowing the catalyst activity and productivity in one slurry loop reactor, what will the catalyst productivity be in a second loop reactor, operated in series with the first one? The following reasoning, logical as it seems, could be adopted, though it should be stressed that it has never been verified experimentally.

Let the catalyst activity profile in the first polymerisation loop reactor be, equation (8.4),

$$\overline{A}_1(t_1, Rx_1) = b_1 \mathrm{e}^{-a_1 t_1} \,.$$

The same catalyst will continue its life in the second reactor. It owns the same type of activity profile there, though the coefficients a and b may be different, due to different reactor conditions,

$$\overline{A}_2(t_2, Rx_2) = b_2 \mathrm{e}^{-a_2 t_2} \,.$$

It would be illogical to assume that the catalyst will start its activity curve all anew in the second reactor, as the particle has already amassed an amount of polymer that will inhibit the polymerisation to a certain extent. The idea from Section 8.3, that the activity of a catalyst particle is decreasing linearly with its own particle productivity may be helpful to solve this difficulty. It is supposed that the catalyst particle starts its activity in the second reactor at the particle productivity $\mathcal{P}_1(\theta_1)$ that was reached when leaving the first reactor, and follows the trajectory of the activity profile in the second reactor from a lifetime $t_{2,0}$ on, that accords with that particle productivity $\mathcal{P}_1(\theta_1)$. This lifetime $t_{2,0}$ can then be calculated,

$$\mathcal{P}_1(\theta_1) = \mathcal{P}_2(t_{2,0}) \,,$$

or,

$$\frac{b_1}{a_1}(1 - \mathrm{e}^{-a_1\theta_1}) = \frac{b_2}{a_2}(1 - \mathrm{e}^{-a_2 t_{2,0}}) \,,$$

whence,

$$t_{2,0} = \frac{-1}{a_2} \log\left[1 - \frac{a_2 b_1}{a_1 b_2}\left(1 - \mathrm{e}^{-a_1\theta_1}\right)\right] . \tag{14.6}$$

The particle productivity in the second reactor is thus generated by following the activity profile $\overline{\mathcal{A}}_2$ as of the time $t_{2,0}$, or, in other words, following the activity profile $\overline{\mathcal{A}}_2$ with its time origin shifted to the later time $t_{2,0}$,

$$\mathcal{P}_2 = \int_{t_2=0}^{\theta_2} b_2 \mathrm{e}^{-a_2(t_2+t_{2,0})} dt_2 \,.$$

Whence,

$$\mathcal{P}_2 = \frac{b_2}{a_2}\mathrm{e}^{-a_2 t_{2,0}}\left(1 - \mathrm{e}^{-a_2\theta_2}\right) ,$$

wherefrom it is visible that the productivity in the second reactor is reduced by a factor $\mathrm{e}^{-a_2 t_{2,0}}$ due to the polymer amassed on the particle in the first reactor. And, after bringing in equation (14.6),

$$\mathcal{P}_2 = \left(\frac{a_1 b_2 - a_2 b_1}{a_1 a_2} + \frac{b_1}{a_1}\mathrm{e}^{-a_1\theta_1}\right)\left(1 - \mathrm{e}^{-a_2\theta_2}\right) . \tag{14.7}$$

Each catalyst fraction that leaves the second reactor is characterised by two residence times, θ_1 and θ_2, in the first and the second reactor respectively. Such a fraction is equal to,

$$E_{2,1}(\theta_1)\mathrm{d}\theta_1 E_{2,2}(\theta_2)\mathrm{d}\theta_2 = \frac{1}{\tau_1\tau_2}\mathrm{e}^{-\theta_1/\tau_1}\mathrm{e}^{-\theta_2/\tau_2}\mathrm{d}\theta_1\mathrm{d}\theta_2 \,.$$

The *catalyst reactor productivity in the second reactor*, $\mathbb{P}_2$, is the sum of the particle productivities of all particles with residence times θ_1 and θ_2,

$$\mathbb{P}_2 = \int_{\theta_1=0}^{\infty}\int_{\theta_2=0}^{\infty} E_{2,1}(\theta_1)\mathrm{d}\theta_1 E_{2,2}(\theta_2)\mathrm{d}\theta_2 \mathcal{P}_2(\theta_1, \theta_2) \,,$$

or,

$$\mathbb{P}_2 = \int_{\theta_1=0}^{\infty}\int_{\theta_2=0}^{\infty} \frac{1}{\tau_1\tau_2}\mathrm{e}^{-\theta_1/\tau_1}\mathrm{e}^{-\theta_2/\tau_2}\left(\frac{a_1 b_2 - a_2 b_1}{a_1 a_2} + \frac{b_1}{a_1}\mathrm{e}^{-a_1\theta_1}\right)\left(1 - \mathrm{e}^{-a_2\theta_2}\right)\mathrm{d}\theta_1\mathrm{d}\theta_2 \,,$$

which is solved as,

$$\mathbb{P}_2 = \frac{b_2 + (a_1 b_2 - a_2 b_1)\tau_1}{1 + a_1\tau_1} \frac{\tau_2}{1 + a_2\tau_2} . \tag{14.8}$$

Equation (14.8), together with equation (8.3),

$$\mathbb{P}_1 = b_1 \frac{\tau_1}{1 + a_1\tau_1} ,$$

could be used to derive the activity profiles in the two reactors from a test run with constant reactor conditions, but different residence times, similar to the tests described in Section 8.4, but executed simultaneously at the two reactors in series. It should be said, however, that such a test has never been carried out by the author.

14.5 The Bimodality Distribution for Two Loop Reactors in Series

As already mentioned, the idea that, in a bimodal production in two loop reactors in series, the polymers of the two different MWD's are most intimately mixed, because both are formed on each catalyst particle, is rather illusive. Due to the nature of the residence time distribution in loop reactors, most particles will carry a polymer that is predominantly formed in the one or the other reactor.

In this field the *bimodality of a polymer particle*, φ, is defined as the fraction of polymer on a catalyst particle that has been formed in the first reactor, whereby it is understood that the remainder of the polymer is produced in the second reactor.
The *split*, Φ, is defined as the fraction of the total polymer production that is produced in the first loop reactor,

$$\Phi = \frac{P_1}{P_1 + P_2} = \frac{\mathbb{P}_1 F_{cat}}{(\mathbb{P}_1 + \mathbb{P}_2)F_{cat}} = \frac{\mathbb{P}_1}{\mathbb{P}_1 + \mathbb{P}_2} .$$

The choice of defining bimodality and split in terms of the first reactor is of course arbitrary. The second reactor is preferred when more importance is attached to the polymer which is produced there, e.g. when that polymer owns the highest molecular weight. Bimodality and split based on the second reactor are equal to $(1-\varphi)$ and $(1-\Phi)$ respectively.

As the point of interest is to know what the bimodality is like for all particles, the *bimodality distribution*, $E(\varphi)$, is defined as a function of the bimodality φ such that $E(\varphi)\mathrm{d}\varphi$ is equal to the weight fraction of the final product of the two reactors with a bimodality between φ and $(\varphi + \mathrm{d}\varphi)$.
The perfect bimodality distribution would be a Dirac delta function $E(\varphi) = \delta(\varphi - \Phi)$ which shows an infinite peak at $\varphi = \Phi$ and is zero for all other values of φ, or, in simple terms, all particles owning a bimodality equal to Φ.
The perfect split is often seen as $\Phi = 0.50$, where the two reactors are running at the same production rate. This split is difficult to maintain as the high and low molecular

weight polymers are difficult to homogenise in the extrusion step of the process, and easily leave gels in the final product, due to drops of high molecular weight polymer that do not dissolve in the matrix of the melt. Usually a compromise is necessary to avoid too high gels levels in the bimodal product. Either the split is set at a different value from 0.50, thus giving more weight to the product from one reactor, or the difference in molecular weight between the two reactor products is reduced, thus producing a product that is less broad or less "bimodal".
In reality the bimodality distribution is far from perfect, and at a split of $\Phi = 0.50$, the distribution is tending to a constant function $E(\varphi) = 1.00$, meaning that all values of bimodality are equally represented in the final polymer. This is shown in the following calculations of the bimodality distribution.

14.5.1 The Ideal Bimodality Distribution

A bimodality distribution can be calculated for a highly simplified case, where, apart from the assumptions already made in this section, is also supposed that,
- the catalyst activity is constant in each reactor, and independent of the residence time.

This *ideal bimodality distribution* is quite easy to calculate and very apt to demonstrate the principle and the nature of the bimodality distribution curves.

Let $\overline{A}_1 = A_1$ be the constant catalyst activity in the first reactor, and $\overline{A}_2 = A_2$ the constant catalyst activity in the second reactor.
The bimodality of a polymer particle formed by a catalyst particle with residence times θ_1 and θ_2 in the first and second reactor respectively, is,

$$\varphi = \frac{A_1\theta_1}{A_1\theta_1 + A_2\theta_2} . \tag{14.9}$$

Consider a certain quantity of catalyst $F_{\text{cat}}\mathrm{d}t$ fed to the first reactor, and passing through the two reactors in series.
The fraction of that catalyst with a residence time θ_2 in the second reactor will be,

$$F_{\text{cat}}\mathrm{d}tE_{2,2}(\theta_2)\mathrm{d}\theta_2 .$$

A subfraction of that fraction will yield a polymer of bimodality φ, if its residence time θ_1 in the first reactor was such that, equation (14.9),

$$\varphi = \frac{A_1\theta_1}{A_1\theta_1 + A_2\theta_2} ,$$

or,

$$\theta_1 = \frac{A_2}{A_1}\frac{\varphi}{1-\varphi}\theta_2 . \tag{14.10}$$

Hence also the relation between $d\theta_1$ and $d\varphi$ for two values θ_2 and θ_1 that match a bimodality value φ, and where θ_2 is arbitrarily chosen and thus independent of φ,

$$d\theta_1 = \frac{A_2}{A_1}\frac{\theta_2}{(1-\varphi)^2}d\varphi\,. \tag{14.11}$$

Bringing in equations (14.10) and (14.11), that subfraction with bimodality φ will be equal to,

$$F_{\text{cat}}dtE_{2,2}d\theta_2E_{2,1}\left(\theta_1 = \frac{A_2}{A_1}\frac{\varphi}{1-\varphi}\theta_2\right)d\theta_1\,,$$

or, expressed in terms of θ_2 only,

$$= F_{\text{cat}}dtE_{2,2}d\theta_2E_{2,1}\left(\theta_1 = \frac{A_2}{A_1}\frac{\varphi}{1-\varphi}\theta_2\right)\frac{A_2}{A_1}\frac{\theta_2}{(1-\varphi)^2}d\varphi\,.$$

To obtain the quantity of polymer produced by this subfraction of catalyst, it is multiplied by the total catalyst particle productivity, also, with the help of equation (14.9), expressed in terms of θ_2,

$$\mathcal{P}_1 + \mathcal{P}_2 = A_1\theta_1 + A_2\theta_2 = \frac{A_2\theta_2}{1-\varphi}\,.$$

The subfraction of polymer with bimodality φ produced from catalyst with a residence time θ_2 in the first reactor is thus,

$$= F_{\text{cat}}dtE_{2,2}d\theta_2E_{2,1}\left(\theta_1 = \frac{A_2}{A_1}\frac{\varphi}{1-\varphi}\theta_2\right)\frac{A_2^2}{A_1}\frac{\theta_2^2}{(1-\varphi)^3}d\varphi\,.$$

To obtain the total fraction of polymer with bimodality φ, this subfraction is integrated over all values of θ_2,

$$= \int_{\theta_2=0}^{\infty} F_{\text{cat}}dtE_{2,2}d\theta_2E_{2,1}\left(\theta_1 = \frac{A_2}{A_1}\frac{\varphi}{1-\varphi}\theta_2\right)\frac{A_2^2}{A_1}\frac{\theta_2^2}{(1-\varphi)^3}d\varphi\,. \tag{14.12}$$

This same total fraction of polymer with bimodality φ can also be expressed in terms of the bimodality distribution function $E(\varphi)$,

$$= F_{\text{cat}}dt(\mathbb{P}_1 + \mathbb{P}_2)E(\varphi)d\varphi = F_{\text{cat}}dt(A_1\tau_1 + A_2\tau_2)E(\varphi)d\varphi\,. \tag{14.13}$$

Equating the expressions (14.12) and (14.13), an equation is obtained wherefrom the bimodality distribution $E(\varphi)$ can be calculated,

$$\begin{aligned} &F_{\text{cat}}dt(A_1\tau_1 + A_2\tau_2)E(\varphi)d\varphi \\ &= \int_{\theta_2=0}^{\infty} F_{\text{cat}}dtE_{2,2}d\theta_2E_{2,1}\left(\theta_1 = \frac{A_2}{A_1}\frac{\varphi}{1-\varphi}\theta_2\right)\frac{A_2^2}{A_1}\frac{\theta_2^2}{(1-\varphi)^3}d\varphi\,, \end{aligned}$$

or, bringing in the residence time distributions,

$$E_{2,2}(\theta_2) = \frac{1}{\tau_2}\exp\left(-\frac{\theta_2}{\tau_2}\right),$$

$$E_{2,1}(\theta_1) = \frac{1}{\tau_1}\exp\left(-\frac{\theta_1}{\tau_1}\right) = \frac{1}{\tau_1}\exp\left(-\frac{A_2}{A_1\tau_1}\frac{\varphi}{1-\varphi}\theta_2\right),$$

$$E(\varphi) = \frac{1}{A_1\tau_1 + A_2\tau_2}\int_{\theta_2=0}^{\infty}\frac{1}{\tau_1\tau_2}\exp\left(-\frac{\theta_2}{\tau_2} - \frac{A_2}{A_1\tau_1}\frac{\varphi}{1-\varphi}\theta_2\right)\frac{A_2^2}{A_1}\frac{\theta_2^2}{(1-\varphi)^3}\mathrm{d}\theta_2\,. \tag{14.14}$$

Finally, the bimodality distribution $E(\varphi)$ for the simplified case of constant catalyst activities is calculated from (14.14) as,

$$E(\varphi) = \frac{2A_1^2A_2^2\tau_1^2\tau_2^2}{(A_1\tau_1 + A_2\tau_2)[A_1\tau_1(1-\varphi) + A_2\tau_2\varphi]^3}\,. \tag{14.15}$$

As expected, this expression is symmetric in A_1, τ_1, φ and $A_2, \tau_2, (1-\varphi)$, its integral over the interval $[\varphi = 0, \varphi = 1]$ is 1, and the average value of all φ values is equal to the split Φ.

By repeated substitution of,

$$\frac{A_1\tau_1}{A_1\tau_1 + A_2\tau_2} = \Phi\,,$$

the ideal bimodality distribution $E(\varphi)$ is found to be a function of the split only,

$$E(\varphi) = \frac{2\Phi^2(1-\Phi)^2}{[\varphi + \Phi(1-2\varphi)]^3}\,. \tag{14.16}$$

In the case where $\Phi = 0.50$ it is easily seen from (14.16) that $E(\varphi) = 1$ for all values of φ.

In Figure 14.1, the ideal bimodality distribution is plotted for three values of the split, $\Phi = 0.40, 0.50$, and 0.60. This distribution is disappointing if a 50/50% distribution of the polymers were expected in each polymer particle. In the best case, at $\Phi = 0.50$, $E(\varphi) = 1$ and all compositions are equally represented. This means that the chance to find a particle composed of pure polymer from the one or the other reactor is as great as the chance to find a particle of a 50/50 or any other composition. When the split is deviating from 0.50, the result gets worse and polymer either from the first or the second reactor is predominant. This is not surprising, as the bimodality is created by the multiplication of two residence time distributions $E_{2,1}(\theta_1)$ and $E_{2,2}(\theta_2)$, where each residence time in the first reactor can be combined with each residence time in the second reactor.

14.5.2 A Realistic Bimodality Distribution

A more detailed, or *realistic* bimodality distribution will now be calculated. The reasoning followed is the same as for the ideal bimodality distribution, only, instead of

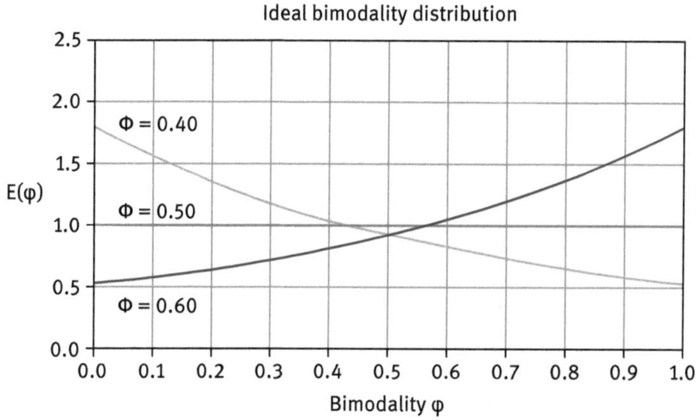

Fig. 14.1. The Ideal Bimodality Distribution for values of the split $\Phi = 0.40, 0.50, 0.60$.

constant catalyst activities, the realistic catalyst productivities from Section 14.4 are used. It will be seen that, in case of equal catalyst activity profiles in both reactors, the results are not so different from the ideal bimodality distribution, and thus that the conclusions are very much the same.

The bimodality of a polymer particle is determined by the ratio of the quantities of the two polymers formed on the particle, and thus by the particle productivities in the first and second reactor,

$$\varphi = \frac{\mathcal{P}_1}{\mathcal{P}_1 + \mathcal{P}_2} .$$

Those particle productivities have been calculated in Sections 8.3 and 14.4, equations (8.6) and (14.7),

$$\varphi = \frac{\frac{b_1}{a_1}(1 - \mathrm{e}^{-a_1\theta_1})}{\frac{b_1}{a_1}(1 - \mathrm{e}^{-a_1\theta_1}) + \left(\frac{a_1 b_2 - a_2 b_1}{a_1 a_2} + \frac{b_1}{a_1}\mathrm{e}^{-a_1\theta_1}\right)(1 - \mathrm{e}^{-a_2\theta_2})} .$$

It is convenient to introduce two *dimensionless catalyst residence times*,

$$K_1 = (1 - \mathrm{e}^{-a_1\theta_1}), K_2 = (1 - \mathrm{e}^{-a_2\theta_2}) , \tag{14.17}$$

in the first and second loop reactors. If θ_i will vary from 0 to $+\infty$, K_i will do so from 0 to 1 ($i = 1, 2$).
Also,

$$d\theta_i = \frac{1}{a_i(1 - K_i)}\mathrm{d}K_i , \tag{14.18}$$

and the residence time distributions in the two reactors, equations (7.9) and (14.3), are,

$$E_{2,i}(\theta_i) = \frac{1}{\tau_i}\exp\left(\frac{-\theta_i}{\tau_i}\right) = \frac{1}{\tau_i}(1 - K_i)^{\frac{1}{a_i\tau_i}} . \tag{14.19}$$

The bimodality φ, expressed in terms of K_i, will be,

$$\varphi = \frac{a_2 b_1 K_1}{a_2 b_1 K_1 + a_1 b_2 K_2 - a_2 b_1 K_1 K_2} . \tag{14.20}$$

Considering a fraction of catalyst with an arbitrary residence time K_2 in the second reactor, that has produced a polymer with a bimodality φ, it follows from equation (14.20) that the residence time in the first reactor must have been,

$$K_1^* = \frac{a_1 b_2}{a_2 b_1} \frac{K_2 \varphi}{1 + \varphi(K_2 - 1)} . \tag{14.21}$$

If $K_2 = 0$, φ will be 1, and K_1^* may have all values between 0 and 1.
If $\varphi = 0$, then K_1^* will be $= 0$ and K_2 may have all values between 0 and 1.
In all other cases, $K_2 > 0$ and $\varphi > 0$, there will be a value of K_1^* that will satisfy a set of values for φ and K_2. Then from equation (14.21), for an arbitrary value of K_2,

$$\mathrm{d}K_1^* = \frac{a_1 b_2}{a_2 b_1} \frac{K_2}{[1 + \varphi(K_2 - 1)]^2} \mathrm{d}\varphi . \tag{14.22}$$

Consider a certain quantity of catalyst $F_{\mathrm{cat}} \mathrm{d}t$ fed to the first reactor.
The fraction of that catalyst with a residence time θ_2, or K_2, in the second reactor will be $E_{2,2}(\theta_2)\mathrm{d}\theta_2$.
A subfraction of that fraction will yield a polymer of bimodality φ, if its residence time in the first reactor is θ_1^* which is equivalent with K_1^*, which satisfies equation (14.21). That subfraction with bimodality φ will be equal to, with the help of equations (14.19) and (14.18),

$$E_{2,2}(\theta_2)\mathrm{d}\theta_2 E_{2,1}(\theta_1^*)\mathrm{d}\theta_1^* = \frac{1}{a_1 a_2 \tau_1 \tau_2} (1 - K_2)^{\frac{1-a_2\tau_2}{a_2\tau_2}} (1 - K_1^*)^{\frac{1-a_1\tau_1}{a_1\tau_1}} \mathrm{d}K_2 \mathrm{d}K_1^* ,$$

or, expressed in terms of φ and K_2 only, by using equations (14.21) and (14.22),

$$= \frac{b_2 K_2}{a_2^2 b_1 \tau_1 \tau_2} (1 - K_2)^{\frac{1-a_2\tau_2}{a_2\tau_2}} \left(1 - \varphi + \frac{a_2 b_1 - a_1 b_2}{a_2 b_1} \varphi K_2\right)^{\frac{1-a_1\tau_1}{a_1\tau_1}} [1 + \varphi(K_2 - 1)]^{\frac{-1-a_1\tau_1}{a_1\tau_1}} \mathrm{d}\varphi \mathrm{d}K_2 .$$

To obtain the quantity of polymer produced by this subfraction of catalyst, it is multiplied by the total catalyst particle productivity, $\mathcal{P}_1(\theta_1^*) + \mathcal{P}_2(\theta_1^*, \theta_2)$.

$$\mathcal{P}_1(\theta_1^*) = \frac{b_1}{a_1}(1 - \mathrm{e}^{-a_1 \theta_1^*}) = \frac{b_1 K_1^*}{a_1} .$$

By definition of φ,

$$\mathcal{P}_1(\theta_1^*) + \mathcal{P}_2(\theta_1^*, \theta_2) = \frac{\mathcal{P}_1(\theta_1^*)}{\varphi} = \frac{b_1 K_1^*}{a_1 \varphi} ,$$

and, by equation (14.21),

$$\mathcal{P}_1(\theta_1^*) + \mathcal{P}_2(\theta_1^*, \theta_2) = \frac{b_2}{a_2} \frac{K_2}{1 + \varphi(K_2 - 1)} \, .$$

The subfraction of polymer with bimodality φ produced from catalyst with a residence time θ_2 or K_2 in the second reactor is thus,

$$= \frac{b_2^2 K_2^2}{a_2^3 b_1 \tau_1 \tau_2} (1-K_2)^{\frac{1-a_2\tau_2}{a_2\tau_2}} \left(1-\varphi + \frac{a_2 b_1 - a_1 b_2}{a_2 b_1} \varphi K_2\right)^{\frac{1-a_1\tau_1}{a_1\tau_1}} [1+\varphi(K_2-1)]^{\frac{-1-2a_1\tau_1}{a_1\tau_1}} \mathrm{d}\varphi \mathrm{d}K_2 .$$

To obtain the total fraction of polymer with bimodality φ this subfraction is integrated over all values of K_2,

$$= \int\limits_{K_2=0}^{1} \frac{b_2^2 K_2^2}{a_2^3 b_1 \tau_1 \tau_2} (1-K_2)^{\frac{1-a_2\tau_2}{a_2\tau_2}} \left(1-\varphi + \frac{a_2 b_1 - a_1 b_2}{a_2 b_1} \varphi K_2\right)^{\frac{1-a_1\tau_1}{a_1\tau_1}} \ldots$$

$$\ldots [1+\varphi(K_2-1)]^{\frac{-1-2a_1\tau_1}{a_1\tau_1}} \mathrm{d}\varphi \mathrm{d}K_2 \, . \tag{14.23}$$

This same fraction of polymer with bimodality φ can also be expressed in terms of the bimodality distribution function $E(\varphi)$,

$$= (\mathbb{P}_1 + \mathbb{P}_2) E(\varphi) \mathrm{d}\varphi \, , \tag{14.24}$$

where, equations (8.3) and (14.8),

$$\mathbb{P}_1 + \mathbb{P}_2 = b_1 \frac{\tau_1}{1 + a_1\tau_1} + \frac{b_2 + (a_1 b_2 - a_2 b_1)\tau_1}{1 + a_1\tau_1} \frac{\tau_2}{1 + a_2\tau_2} = \frac{b_1\tau_1 + b_2\tau_2 + a_1 b_2 \tau_1 \tau_2}{(1 + a_1\tau_1)(1 + a_2\tau_2)} \, .$$

Equating the expressions (14.23) and (14.24), the bimodality distribution $E(\varphi)$ is obtained,

$$E(\varphi) = \frac{(1 + a_1\tau_1)(1 + a_2\tau_2)}{b_1\tau_1 + b_2\tau_2 + a_1 b_2 \tau_1 \tau_2} \int\limits_{K_2=0}^{1} \frac{b_2^2 K_2^2}{a_2^3 b_1 \tau_1 \tau_2} (1-K_2)^{\frac{1-a_2\tau_2}{a_2\tau_2}} \ldots$$

$$\ldots \left(1-\varphi + \frac{a_2 b_1 - a_1 b_2}{a_2 b_1} \varphi K_2\right)^{\frac{1-a_1\tau_1}{a_1\tau_1}} [1+\varphi(K_2-1)]^{\frac{-1-2a_1\tau_1}{a_1\tau_1}} \mathrm{d}K_2 \, . \tag{14.25}$$

Equation (14.25) allows to calculate all possible bimodality distributions $E(\varphi)$ for all catalyst activity profiles $\overline{A}_1(t_1, Rx_1) = b_1 \mathrm{e}^{-a_1 t_1}$ and $\overline{A}_2(t_2, Rx_2) = b_2 \mathrm{e}^{-a_2 t_2}$ in the first reactor and second reactor respectively, at the residence times τ_1 and τ_2, by a numerical solution of the integral.

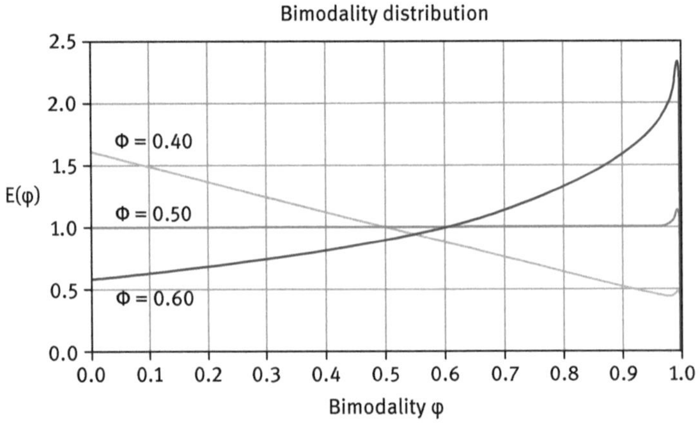

Fig. 14.2. Realistic Bimodality Distribution for a Ziegler–Natta catalyst with productivity parameters $a = 0.398$ and $b = 20\,926$ in both reactors. The residence times were adapted so that splits $\Phi = 0.40, 0.50, 0.60$, were obtained.

Three examples are shown in Figure 14.2 for a Ziegler–Natta catalyst, assuming that the activity profiles in the two reactors are identical. The residence times were adapted so that splits $\Phi = 0.40$, 0.50, and 0.60 were obtained. The choice of identical activity profiles in the two reactors is not producing the most spectacular curves, but is realistic because, in a bimodal process, catalyst designers will strive for identical catalyst activities in both reactors, hoping to obtain a well balanced bimodal product. When comparing the results for the ideal and the realistic bimodality distributions in Figures 14.1 and 14.2, it is clear that the two distributions are very similar if the conditions assumed for the realistic distribution are close to the assumptions for the ideal distribution.

The realistic curves, however, are not symmetric. They show a peak at values of $\varphi \to 1$ before dropping to zero, because catalyst particles with long residence times in the first reactor will have lost most of their activity when entering the second reactor, due to the polymer amassed on the particle, and thus will tend to high values of φ. This fact is not taken into account by the ideal bimodality distribution, where all particles in the second reactor are supposed to own the same constant activity, but it should be considered when deciding which polymer (high or low molecular weight) will be produced in the first reactor.

Notations

The reactors are indicated by a subscript:

1	first loop reactor, or reactor 1
2	second loop reactor, or reactor 2

a	constant in the expressions for the catalyst activity profile and catalyst productivity, (s^{-1}).
$\mathbb{A}$	catalyst reactor activity: quantity of polymer produced by all catalyst present in the reactor per unit time, per *kg* catalyst, (*kg* polymer/*kg* catalyst, *s*).
$\overline{\mathcal{A}}$	average activity profile of the catalyst: quantity of polymer produced per *kg* of catalyst, per unit time, at a certain lifetime *t* in the reactor, (*kg* polymer/*kg* catalyst, *s*).
b	constant in the expressions for the catalyst activity profile and catalyst productivity, (*kg*, kg^{-1}, s^{-1}).
$E(\varphi)$	bimodality distribution: a function of the bimodality φ such that $E(\varphi)\mathrm{d}\varphi$ is equal to the weight fraction of the final product of the two reactors with a bimodality between φ and $(\varphi + \mathrm{d}\varphi)$, (–).
$E_{2,1}(\theta_1)$	catalyst residence time distribution in the first reactor: a function of the catalyst residence time θ_1 such that $E_{2,1}(\theta_1)\mathrm{d}\theta_1$ is equal to the weight fraction of the catalyst in the first reactor with a residence time between θ_1 and $(\theta_1 + \mathrm{d}\theta_1)$, ($s^{-1}$).
$E_{2,2}(\theta_2)$	catalyst residence time distribution in the second reactor: a function of the catalyst residence time θ_2 such that $E_{2,2}(\theta_2)\mathrm{d}\theta_2$ is equal to the weight fraction of the catalyst in the second reactor with a residence time between θ_2 and $(\theta_2 + \mathrm{d}\theta_2)$, ($s^{-1}$).
$E_{2,1+2}(\theta)$	total catalyst residence time distribution in two reactors in series: a function of the total catalyst residence time θ in the two reactors such that $E_{2,1+2}(\theta)\mathrm{d}\theta$ is equal to the weight fraction of the catalyst leaving the second reactor with a total residence time in the two reactors between θ and $(\theta + \mathrm{d}\theta)$, ($s^{-1}$).
F_{cat}	catalyst feed to the two reactors, injected in the first reactor, (*kg*, s^{-1}).
K_1	dimensionless catalyst residence time $K_1 = (1 - \mathrm{e}^{-a_1\theta_1})$ in the first loop reactor, (–).
K_2	dimensionless catalyst residence time $K_2 = (1 - \mathrm{e}^{-a_2\theta_2})$ in the second loop reactor, (–).
$m_{\mathrm{cat},2}$	mass of catalyst present in the second reactor, (*kg*).
$m_{\mathrm{pol},2}$	mass of polymer present in the second reactor, (*kg*).
P	polymer production rate in a reactor: the total quantity of polymer produced in the reactor per unit of time, (*kg* polymer, s^{-1}).
$\mathbb{P}$	catalyst reactor productivity: total quantity of polymer produced in the reactor by one *kg* of catalyst that has left the reactor, (*kg* polymer/*kg* catalyst).
$\mathcal{P}$	catalyst particle productivity: quantity of polymer produced by one catalyst particle during its lifetime at a certain concentration profile, (*kg* polymer/*kg* catalyst).

q_0	a small quantity of tracer catalyst, injected at time $\theta_2 = 0$ in the second reactor, (kg).
q_2	quantity of the tracer catalyst q_0, present in the second reactor, (kg).
Rx	symbol representing all of the reactor conditions, including temperature, concentrations, and flows.
t	catalyst lifetime: the time interval between the moment when a catalyst particle that is present in the reactor, has entered the reactor, and any moment at which the reactor system is considered, (s).
θ	total catalyst residence time in two reactors in series, (s).
θ_1	catalyst residence time in the first reactor, (s).
θ_2	catalyst residence time in the second reactor, (s).
τ_1	average catalyst residence time in the first reactor, (s).
τ_2	average catalyst residence time in the second reactor, (s).
τ_{1+2}	total average catalyst residence time in two loop reactors in series, (s).
φ	bimodality of a polymer particle: the fraction of polymer on a catalyst particle that has been formed in the first reactor, (–).
Φ	split: the fraction of the total polymer production that is produced in the first reactor, (–).

Index

Bei Fragen zur Produktsicherheit wenden Sie sich bitte an:
If you have any questions regarding product safety,
please contact:

Walter de Gruyter GmbH
Genthiner Straße 13
10785 Berlin
productsafety@degruyterbrill.com